Reinforcement Learning for Cyber Operations

IEEE Press
445 Hoes Lane
Piscataway, NJ 08854

Reinforcement Learning for Cyber Operations

Applications of Artificial Intelligence for Penetration Testing

Abdul Rahman, PhD
Washington, D.C., USA

Christopher Redino, PhD
New York, New York, USA

Dhruv Nandakumar
Boston, MA, USA

Tyler Cody, PhD
Virginia Tech, USA

Sachin Shetty, PhD
Old Dominion University
Suffolk, Virginia, USA

Dan Radke
Arlington, VA, USA

IEEE Press
WILEY

Published by John Wiley & Sons, Inc., Hoboken, New Jersey.

Published simultaneously in Canada.

Library of Congress Cataloging-in-Publication Data applied for:

Hardback ISBN: 9781394206452

Cover Design: Wiley
Cover Image: © Israel Sebastian/Getty Images

Set in 9.5/12.5pt STIXTwoText by Straive, Chennai, India

To the courageous pioneers in cybersecurity - whose unwavering commitment and bold actions protect our digital frontier, this book is humbly dedicated with the utmost respect and gratitude to helping your cause.

Contents

List of Figures

About the Authors

Dr. Abdul Rahman holds PhDs in physics, math, information technology–cybersecurity and has expertise in cybersecurity, big data, blockchain, and analytics (AI, ML).

Dr. Christopher Redino holds a PhD in theoretical physics and has extensive data science experience in every part of the AI/ML lifecycle.

Mr. Dhruv Nandakumar has extensive data science expertise in deep learning.

Dr. Tyler Cody is an Assistant Research Professor at the Virginia Tech National Security Institute.

Dr. Sachin Shetty is a Professor in the Electrical and Computer Engineering Department at Old Dominion University and the Executive Director of the Center for Secure and Intelligent Critical Systems at the Virginia Modeling, Analysis and Simulation Center.

Mr. Dan Radke is an Information Security professional with extensive experience in both offensive and defensive cybersecurity.

Foreword

I first met Dr. Abdul Rahman when he was serving as research faculty at the Hume Center at Virginia Tech, and I was building an AI Center of Excellence at one of the largest companies in the world. Abdul's energy and creativity are the first thing I noticed, quickly followed by his expertise that has been acquired through his decades of experience as a cybersecurity professional. All of this provides a continuous source of unique insights into where AI/ML (machine learning) can make the biggest impact in protecting organizations from cyber threats. What's been equally impressive to me is his ability to build teams, partner, and collaborate. This book and the group of talented authors assembled to write it are a great example of his ability to build impressive and productive teams.

This work is timely as the nature of cyber threats is evolving at an accelerated pace as attack surfaces continue to expand, nation-state actors become more sophisticated, and ML provides a force multiplier for adversaries. As ML (a subset of artificial intelligence [AI]) becomes more sophisticated, organizations are going to be required to use ML techniques to respond. ML will be a force multiplier for organizations as a small number of operators with AI tools can defend against a higher volume of activities with faster response times. In addition, AI/ML will provide detection capabilities outside the ability of existing rules-based technologies and humans to detect. The norm will be using "AI to combat AI."

Reinforcement learning (RL), as an ML technique, thrives in situations where a physical environment is being modeled. Trial and failure, or reward and penalties, are central to constructing RL models. A network is an excellent environment for applying RL: it's intuitively easy to understand the concept of an agent navigating a network and receiving rewards for remaining undetected and penalties when encountering well-secured nodes.

I've had the pleasure of working with the authors on a number of ML research projects designed to improve the ability of organizations to detect adversarial campaigns. They have published numerous papers in peer-reviewed scientific forums, a few of which I have had the pleasure of coauthoring. The work is industry-leading and has been deployed at commercial, government, and nonprofit organizations. This esteemed group of authors has successfully bridged the gap between academic research and commercial deployment in a powerful way. The contents of this book are not just academic; these techniques have been applied successfully in the real world. As the age of AI continues to evolve, rapidly translating research into practical commercial tools is going to be critical.

The authors have done a wonderful job of telling the story of the research, development, and application journey that we have been on for the last few years. I hope that you enjoy the book and take away a few of the many useful lessons contained within. Happy reading.

Edward Bowen, PhD

Preface

Artificial intelligence (AI)-driven penetration testing using reinforcement learning (RL) to support cyber defense operations (i.e., blue teaming) enables optimizations in workflows to increase visibility, optimize analyst workflows, and improve cyber detections. Similarly, teams dedicated to developing cyber resilience plans through penetration testing (i.e., red–purple–blue teaming) can leverage this type of RL to identify exploitable weaknesses to improve an organization's cyber posture.

Abdul Rahman
Christopher Redino
Dhruv Nandakumar
Tyler Cody
Sachin Shetty
Dan Radke
USA

Acknowledgments

We thank our esteemed professional colleagues. We also thank the fantastic editorial staff at Wiley-IEEE Indirakumari S., Mary Hatcher, and Victoria Bradshaw for their patience, dedication, and professionalism. We thank our reviewers for their comments and recommendations on improving the manuscript. We also thank our families and each other for the effort, time, and dedication required to see this project through. We are grateful for Dr. Edward Bowen's unwavering support in incubating and nurturing our efforts from their inception. Dr. Sachin Shetty encouraged our group at the outset to pursue this manuscript; we are very grateful for his collaboration and support. Special thanks to Irfan Saif, Emily Mossberg, Mike Morris, Will Burns, Eric Dull, Dr. Laura Freeman, Dr. Peter Beling, Adnan Amjad, Deb Golden, Eric Hipkins, Joe Nehila, Joe Tobolski, and Ranjith Raghunath for their encouragement and support.

Dr. Abdul Rahman is grateful for his family, coauthors, colleagues, and the great staff at Wiley for supporting this effort.

Dr. Christopher Redino thank Ren Redino, his own little neutral network that he did not design, but he does try to train.

Dhruv Nandakumar would like to thank his family and friends for their invaluable support through the authorship process. He would also like to thank his incredibly talented co-authors without whom this would have not been possible.

Dr. Tyler Cody is grateful to his wife, Bingxu, and their loyal dog, Meimei.

Dr. Sachin Shetty is grateful for his family and coauthors.

Dan Radke extends his deepest thanks to his incredible wife, Danah, for her love and infinite patience. She is the steadfast anchor to reality and the brilliant spark that fuels his journey to dream big and venture into the wonderfully weird.

Abdul Rahman
Christopher Redino
Dhruv Nandakumar
Tyler Cody
Sachin Shetty
Dan Radke

Acronyms

A2C	advantage actor to critic
AD	active directory
ALE	annualized loss exposure
APT	advanced persistent threat
AV	antivirus
C2	command and control
CISO	chief information security officer
CPE	common platform enumeration
CVE	common vulnerability and exposures
CVSS	common vulnerability scoring system
CWE	common weakness enumeration
DAA	double agent architecture
DLP	data loss prevention
DQN	deep Q network
EDR	end point detection and response
EPP	end point protection
HID	host-based intrusion detection
HIP	host-based intrusion prevention
IAPTS	intelligent automated penetration testing system
IDS	intrusion detection system
IPB	intelligence preparation of the battlefield
IPS	intrusion prevention system
LDAP	lightweight directory access protocol
LEF	loss event frequency
LM	loss magnitude
LRM-RAG	layered reference model for RL with attack graphs
MDP	Markov decision processes
MOL	multi objective learning
MORL	multi objective reinforcement learning

MTL	multi task learning
Nmap	network mapper
OSINT	open source intelligence
PII	personally identifiable information
PPO	proximal policy optimization
PT	penetration testing
RL	reinforcement learning
SDA	situational domain awareness
SDR	surveillance detection routes
SMB	server message block
SNMP	simple network management protocol
TEF	threat event frequency
TTP	tactics, techniques, and procedures
YAML	YAML ain't markup language

Introduction

This book focuses on using reinforcement learning (RL), a type of artificial intelligence (AI), to learn about the traversal of optimal network paths to identify valuable operational details for cyber analysts. Bad actor intent inspires network path traversal leading toward adversarial campaign goals. Among an adversary's key objectives is moving through their target network while minimizing detection. Cyber defenders strive to optimize sensor placement in key network locations and to optimize the visibility facilitating the detection of hackers, threats, and detection of malicious software (malware)/threats. The AI proposed in this book is designed to accomplish both goals.

A network consists of devices (switches, routers, firewalls, intrusion detection systems, intrusion prevention systems, endpoints, and various security appliances) deployed to maximize protection for the organization's information technology (IT) resources. Anticipating vulnerabilities and potential exploits to enhance cyber defense, commonly called the "blue picture," typically requires analyzing collected data from sensors and agents across diverse networks. Understanding patterns based on observed network activities involves careful curation and processing of this data to facilitate developing, training, testing, and validating AI / Machine Learning (ML) / analytics models for detecting nefarious behaviors. Although this traditional approach has yielded some results in the past, this book aims to present a different methodology for pinpointing weaknesses in networks. This book discusses how to translate physical networks into logical representations called attack graphs and employ RL AI models to predict vulnerabilities, optimal visibility, and weaknesses within these network topologies.

Adversarial attack campaigns vary by intention, from establishing a foothold to full disruption. A core premise in most advanced persistent threat (APT) and other sophisticated adversary's objectives is to develop an approach to understanding an environment (hosts, devices, network, etc.) to include their roles and relative positioning of relevant devices in the network. Adversaries focus on developing "connectivity maps" while developing attack campaigns to inform, shape, and drive clear focus on their goals. Among these goals is learning about the environment via undetected reconnaissance methods to ultimately map out the adversarial attack paths, systems, and applications leading to the compromise of these key assets (also called crown jewels) or theft of data (also called exfiltration) within them.

In this book, the physical network has to be translated into a logical structure called an attack graph [2] so the RL AI can provide useful predictions. In RL, a reward is attributed to the detection value of devices in the network including their inherent "cyber terrain" characterization [1]. The RL AI inspired from adversarial attack campaigns presented in this book motivates different reward systems, the network's topology, and cyber terrain informed by likely key cyber campaign objectives. Examples of these objectives consist of optimal paths to exfiltrate data or the best location in the network to perform reconnaissance without being append "(e.g. surveillance detection route)". In this book, we will discuss the requirements for building the RL AI to support deeper and broader coverage within the book, along with several cyber-specific use cases that are useful for cyber analysts, threat researchers, cyber hunters, and cyber operations teams.

Traditional network penetration tests are one of the best ways to turn these network representations into network protection realities. Specialized teams perform these tests teams with highly technical expertise. This expertise includes software design, network transmission protocols and standards, and how businesses and private persons use computers. The human element of the testing is irreplaceable, the variables of the vulnerabilities are complex, and no network is like any other. These complexities require using various technologies, tools, and applied knowledge during a test. However, the results of the tests are often acquired through manual and tedious means due to the overabundance of information, logs, and network endpoints that must be sufficiently aggregated to hunt for vulnerabilities. Using automated tooling to collect information is a concept that has been introduced previously. However, network defenders and red teams still face the significant challenge of analyzing and post-collection intelligence to enable "sense-making". However, AI capabilities have drastically improved over the past five years. The RL AI model approach proposed in this book can be used to sift through mountains of previously ignored data to find patterns, anomalies, and graph-linked data epiphanies helping analysts and operators accelerate their ability to thwart bad actors with malicious intent.

References

1 Greg Conti and David Raymond. *On cyber: towards an operational art for cyber conflict*. Kopidion Press, 2018.

2 Xinming Ou, Wayne F Boyer, and Miles A McQueen. A scalable approach to attack graph generation. In *Proceedings of the 13th ACM Conference on Computer and Communications Security*, CCS '06, pages 336345, New York, NY, USA, 2006. Association for Computing Machinery.

1

Motivation

1.1 Introduction

Reinforcement learning (RL) applied to penetration testing has demonstrated feasibility, especially when considering constraints on the representation of attack graphs, such as scale and observability [8]. As adversaries build high-fidelity maps of target networks through reconnaissance methods that populate network topology pictures, a deeper understanding of optimal paths to conduct (cyber) operations becomes clear. In this respect, cyber defenders (blue teams) and cyberattackers (red teams) employ principles of visibility optimization to improve or diminish detection efficacy. Protecting/exploiting targets within complex network topologies involves understanding that detection evasion involves keenly traversing paths by reducing visibility. On the other hand, knowledge of such paths empowers blue teams (cyber defenders) by identifying "blind spots" and weaknesses in the network where defenses can be improved. This book is motivated by the prevailing belief that an attack campaign is a well-planned orchestration by well-equipped adversaries involving keen traversal of network paths to avoid detection. Figure 1.1 depicts the "bookends" of the MITRE ATT&CK framework as the focus on tactics that can be detected through network detection (ND). Whereas the "internal" part of the framework focuses on tactics that can be detected through end-point (EP) detection, the "book ends" typically entail network-centric detections. In this methodology, artificial intelligence (AI) models can learn through direct interaction with attack graphs enriched with scan data derived from EP and network information rather than relying on a fixed and carefully curated dataset. This book aims to offer "blue team" and "red team" operations staff the ability to utilize the RL AI methods developed in this book

Reinforcement Learning for Cyber Operations: Applications of Artificial Intelligence for Penetration Testing, First Edition. Abdul Rahman, Christopher Redino, Dhruv Nandakumar, Tyler Cody, Sachin Shetty, and Dan Radke.

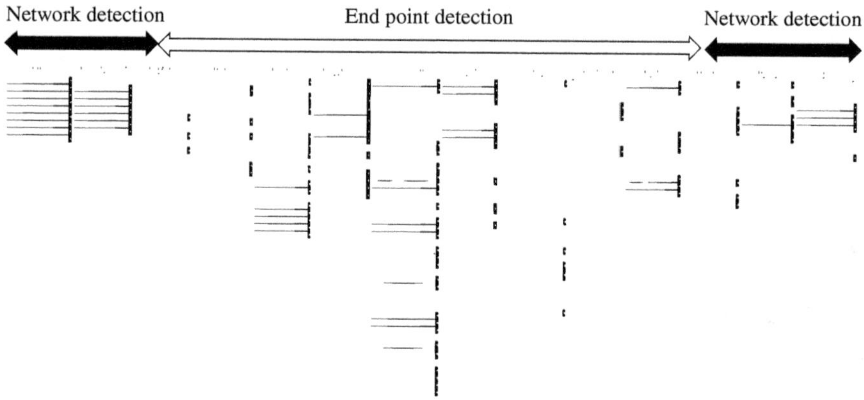

Figure 1.1 MITRE ATT&CK framework identifying the tactics for both network and end-point detection.

to improve visibility, optimize operations, clarify cyber planning objectives, and improve overall cybersecurity posture.

This chapter introduces the key technical elements that provide a foundation for the use of AI. MITRE's Fort Meade Experiment (FMX) research environment [25] was used to systematically categorize adversary behavior during structured emulation exercises within test networks. In 2010, FMX served as a dynamic testing ground, providing researchers with a "living lab" capability to deploy tools and refine ideas for more effective threat detection within the MITRE corporate network [25]. MITRE initiated research within FMX to expedite the detection of advanced persistent threats (APTs) under an "assume breach" mindset. Periodic cyber game exercises emulated adversaries in a closely monitored environment, while threat hunting tested analytic hypotheses against collected data, all with the overarching goal of enhancing post-compromise threat detection in enterprise networks through telemetry sensing and behavioral analytics [1, 16, 25].

ATT&CK played a central role in the FMX research, initially crafted in September 2013 with a primary focus on the Windows enterprise environment. Over time, it underwent refinement through internal research and development, leading to its public release in May 2015 with 96 techniques organized under 9 tactics. After its release, ATT&CK experienced substantial growth, fueled by contributions from the cybersecurity community. MITRE introduced additional ATT&CK-based models, expanding beyond Windows to include Mac and Linux in 2017 (ATT&CK for Enterprise). Other models, such as PRE-ATT&CK (2017), ATT&CK for Mobile (2017), ATT&CK for Cloud (2019), and ATT&CK for ICS (2020), addressed specific domains [16, 25].

ATT&CK, functioning as a knowledge base for cyber adversary behavior and taxonomy, consists of two parts: ATT&CK for Enterprise (covering behavior against enterprise IT networks and cloud) and ATT&CK for Mobile (focusing on behavior against mobile devices). Its inception in 2013 aimed to document common tactics, techniques, and procedures (TTPs) employed by APTs on Windows enterprise networks within the context of the FMX research project.

The framework's significance lies in documentation and in providing behavioral observables for detecting attacks by analyzing cyber artifacts. It employs the structure of TTP to help analysts understand adversarial actions, organize procedures, and fortify defenses. Despite highlighting numerous techniques, ATT&CK falls short in offering insights into how adversaries combine techniques, emphasizing the need for well-defined technique associations for constructing TTP chains. The TTP structure enables analysts to categorize adversarial actions into specific procedures related to particular techniques and tactics, facilitating an understanding of an adversary's objectives and enhancing defense strategies. In addition, these techniques and procedures also serve as indicators of behavior for detecting attacks by scrutinizing cyber artifacts obtained from network and end-system sources [16].

While MITRE ATT&CK comprehensively outlines various techniques an adversary may employ, the necessary associations between techniques for constructing TTP chains remain insufficiently specified. Establishing these associations is crucial as they assist analysts and operators in reasoning about adversarial behavior and predicting unobserved techniques based on those observed in the TTP chain (i.e., unknown behavior). Without well-defined technique associations, cybersecurity professionals face challenges in efficiently navigating the growing search space, especially as the number of TTP chains expands exponentially with the increasing variety of techniques [1]. Given the limited exploration of technique correlations to date, ATT&CK concentrates on acquiring knowledge about the associations between attack techniques, revealing interdependencies, and relationships through analyzing real-life attack data [1, 16, 25].

In classifying reported attacks, the framework distinguishes between APTs and software attacks. APT attacks align with MITRE's threat actor "groups," while software attacks encompass various malicious code forms. Each comprises post-exploit techniques constituting the TTP chain of APTs or software. They use discrete variables, specifically asymmetrical binary variables, with outcomes of 0 or 1 representing negative or positive occurrences of a technique in an attack instance, respectively [1].

The notable limitations outlined by MITRE include a subset representation of techniques and potential biases in heuristic mappings. There is a need to characterize APT and software attacks driven by a requirement to continually evaluate the evolving threat landscape dynamics. MITRE acknowledges a constraint in their data collection process, emphasizing that APT and software attacks may not represent the full spectrum of techniques employed by associated threat actors.

Instead, the framework offers a subset based on publicly available reporting, making it challenging to ascertain the true techniques employed in an adversarial attack campaign (i.e., the definition of operational ground truth). Second, the framework is subject to mapping biases, where heuristics and automated mappings of threat reports to techniques may inadvertently exhibit bias. Recognizing these limitations, MITRE ATT&CK and the processes/workflows to keep it up to date embody an approach for continually refining the characterization of APT and software attacks informing on the scope of possible adversarial TTPs in an attack campaign [1]. While not optimal, to date, this represents one of the few broadly accepted methodologies for characterizing adversarial workflows [25].

1.1.1 Cyberattack Campaigns via MITRE ATT&CK

A cyberattack campaign can follow a workflow captured within the MITRE ATT&CK [16] framework. The workflow progresses from left to right where the left column represents the collection of tactics for preplanning. As Figure 1.1 depicts the "bookends" of the kill chain as being relegated to ND capabilities, the middle portion is typically aligned with activities on EPs. Attack campaigns start with a goal that may involve data exfiltration from key information technology (IT) assets, typically called crown jewels (CJs). These systems, databases, and devices are of high value within an organization. While a large emphasis is placed on detecting both known and unknown threats through using agents deployed on EPs, detection of new low signal-to-noise (STN) sophisticated adversarial attacks evades most detection capabilities. Unfortunately, the constant tension between building better EP detections to trigger indicators of compromise (IOC) from new threats is typically hit or miss [12, 13, 15, 23, 26].

1.2 Attack Graphs

The flaw hypothesis model outlines a general process involving the gathering of information about a system, formulating a list of hypothetical flaws (generated, e.g., through domain expert brainstorming), prioritizing the list, testing hypothesized flaws sequentially, and addressing those that are discovered. McDermott emphasizes the model's applicability to almost all penetration testing scenarios [14]. The attack tree model introduces a tree structure to the information-gathering process, hypothesis generation, etc., offering a standardized approach to manual penetration testing and providing a foundation for automated penetration testing methods [19, 20]. The attack graph model introduces a network structure, distinguishing itself from the attack tree model in terms of the richness of topology, and the corresponding amount of information required to specify the model [7, 14, 18].

Automated penetration testing, integrated into practice [24], relies on the attack tree and attack graph models as their foundation. In RL, these models involve constructing network topologies, treating machines (i.e., servers and network devices) as vertices and links between machines as edges. Variations include additional details about subnetworks and services. In the case of attack trees, probabilities are assigned to branches between parent and child nodes. For attack graphs, transition probabilities between states are assigned to each edge. This is described in more detail in Chapters 3 and 4.

While many advantageous properties of attack trees persist in attack graphs, it remains uncertain whether attack graphs can outperform attack trees in systems that are largely undocumented, i.e., systems with partial observability [14]. RL for penetration testing utilizes the attack graph model, treating the environment either as a Markov decision process (MDP), reflecting classical planning with deterministic actions and known network structure, or as a partially observable Markov decision process (POMDP), where action outcomes are stochastic and network structure and configuration are uncertain [9, 21, 28].

1.3 Cyber Terrain

The foundational concept of the terrain is integral to intelligence preparation of the battlefield (IPB) [5, 10]. In the physical realm, terrain pertains to land and its features. In [5], a definition of cyber terrain is given as "the systems, devices, protocols, data, software, processes, cyber personas, and other network entities that comprise, supervise, and control cyberspace." Cyber terrain emphasizes operations at the strategic, operational, and tactical levels, encompassing elements such as transatlantic cables and satellite constellations, telecommunications offices and regional data centers, and wireless spectrum and local area network protocols. The use of RL engages in the logical plane of cyber terrain, which includes the data link, network, network transport, session, presentation, and application layers (i.e., layers 2–7) of the open systems interconnection (OSI) model [11].

Terrain analysis typically follows the OAKOC framework, consisting of observation and fields of fire (O), avenues of approach (A), key and decisive terrain (K), obstacles (O), and cover and concealment (C[CE2]). These notions from traditional terrain analysis can be applied to cyber terrain [2, 5]. For example, fields of fire may concern all that is network reachable (i.e., line of sight), and avenues of approach may consider network paths inclusive of available bandwidth [5]. In this previous work, we use obstacles to demonstrate how our methodology can be

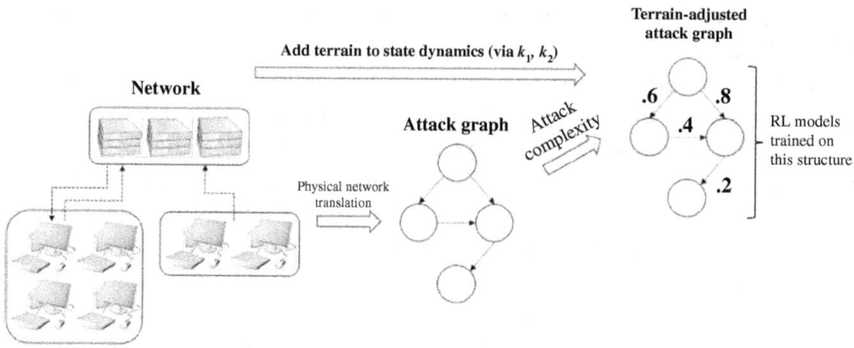

Figure 1.2 Mapping of physical network into an attack graph.

used to bring the first part of the OAKOC framework to attack graph construction for RL [8]. The RL described in this book illustrates how our methodology can integrate the initial aspect of the OAKOC framework into attack graph construction for RL. Cyber terrain functions for each type of device are annotated into the attack graphs prior to the RL running over them (Figure 1.2).

1.4 Penetration Testing

Penetration testing (pen testing) is defined by Denis et al. as, "a simulation of an attack to verify the security of a system or environment to be analyzed... through physical means utilizing hardware, or through social engineering" [6]. They continue by emphasizing that penetration testing is different from port scanning. Specifically, if port scanning is looking through binoculars at a house to identify entry points, penetration testing is having someone break into the house.

Pen testing is part of broader vulnerability detection and analysis, which typically combines penetration testing with static analysis [3, 4, 22]. Penetration testing models have historically taken the form of the flaw hypothesis model [17, 27], the attack tree model [19, 20], or the attack graph model [7, 14, 18]. A detailed discussion of penetration testing is covered in Chapter 2.

1.5 AI Reinforcement Learning Overview

In the sophisticated arena of cybersecurity, the integration of AI and, more specifically, RL heralds a transformative approach to enhancing penetration testing within network systems. This introductory section foreshadows the comprehensive exploration in Chapter 3, setting the stage for an in-depth discussion on the pivotal role of RL in devising more efficient and robust penetration testing methodologies.

RL emerges as a quintessential paradigm for penetration testing, attributed to its inherent adaptability and sophisticated decision-making properties. While conventional machine learning methodologies excel in prediction-based tasks, they often falter in the face of dynamic and unpredictable scenarios characteristic of network security. In contrast, RL excels by learning to formulate strategies through interaction, thereby making it exceptionally suitable for the multifaceted and unpredictable domain of network security.

Central to RL is the concept of an agent that learns to make decisions by interacting with its environment to achieve a defined objective. This paradigm mirrors the process of a penetration tester navigating through a network, making strategic decisions at each juncture to delve deeper while remaining undetected. With each interaction, the agent acquires knowledge, incrementally refining its strategy to maximize efficacy – akin to how a human tester enhances their techniques through experience.

Chapter 3 is dedicated to elucidating the theoretical and mathematical underpinnings of RL. It commences with a delineation of the environment, states, actions, and rewards, progressing to dissect the MDPs. MDPs offer a mathematical framework to model decision-making scenarios where outcomes are influenced by both randomness and the decisions of the agent, resonating deeply with the unpredictable nature of penetration testing.

The discourse will extend to deep reinforcement learning (DRL), highlighting how neural networks are employed to manage high-dimensional inputs and complex policy formulations. The capability of DRL to process and make informed decisions based on extensive and intricate data is indispensable for navigating and exploiting sophisticated network architectures.

As readers embark on the journey through Chapter 3, it is imperative to recognize that the exploration of RL is not merely about comprehending algorithms but about appreciating their transformative potential in redefining penetration testing. The forthcoming chapter will furnish the technical acumen necessary to fully understand these concepts and explore their applicability to real-world security challenges.

The application of RL in penetration testing signifies a substantial advancement, offering a methodology that learns, adapts, and dynamically optimizes strategies. As readers proceed, they should remain cognizant of the potential of these techniques not only to understand network vulnerabilities but also to anticipate and mitigate evolving security threats. Chapter 3 promises to unfold these concepts meticulously, paving the path for a new paradigm of intelligent and autonomous penetration testing.

1.6 Organization of the Book

Chapter 1 introduces the book's focus on the intersection of AI, in particular, RL, with cybersecurity. Chapter 2 discusses current approaches to penetration testing followed by a review of RL in Chapter 3. Chapter 4 discusses the motivation for using RL for penetration testing followed by how to operationalize these RL models in Chapter 5. RL for penetration testing from a practical standpoint is covered in Chapter 6 followed by scaling considerations in Chapter 7. Extending and using these models is covered in Chapter 8 followed by the conclusion in Chapter 9.

References

1 Rawan Al-Shaer, Jonathan M Spring, and Eliana Christou. Learning the associations of MITRE ATT&CK adversarial techniques. In *2020 IEEE Conference on Communications and Network Security (CNS)*, pages 1–9. IEEE, 2020.

2 Scott D Applegate, Christopher L Carpenter, and David C West. Searching for digital hilltops. *Joint Force Quarterly*, 84(1):18–23, 2017.

3 Aileen G Bacudio, Xiaohong Yuan, Bei-Tseng B Chu, and Monique Jones. An overview of penetration testing. *International Journal of Network Security & Its Applications*, 3(6):19, 2011.

4 Brian Chess and Gary McGraw. Static analysis for security. *IEEE Security & Privacy*, 2(6):76–79, 2004.

5 Greg Conti and David Raymond. *On cyber: towards an operational art for cyber conflict*. Kopidion Press, 2018.

6 Matthew Denis, Carlos Zena, and Thaier Hayajneh. Penetration testing: concepts, attack methods, and defense strategies. In *2016 IEEE Long Island Systems, Applications and Technology Conference (LISAT)*, pages 1–6. IEEE, 2016.

7 Bing Duan, Yinqian Zhang, and Dawu Gu. An easy-to-deploy penetration testing platform. In *2008 The 9th International Conference for Young Computer Scientists*, pages 2314–2318. IEEE, 2008.

8 Rohit Gangupantulu, Tyler Cody, Paul Park, Abdul Rahman, Logan Eisenbeiser, Dan Radke, and Ryan Clark. Using cyber terrain in reinforcement learning for penetration testing. In *2022 IEEE International Conference on Omni-layer Intelligent Systems (COINS)*, pages 1–8, 2022.

9 Mohamed C Ghanem and Thomas M Chen. Reinforcement learning for intelligent penetration testing. In *2018 Second World Conference on Smart Trends in Systems, Security and Sustainability (WorldS4)*, pages 185–192. IEEE, 2018.

10 Jeffrey Guion and Mark Reith. Cyber terrain mission mapping: tools and methodologies. In *2017 International Conference on Cyber Conflict (CyCon U.S.)*, pages 105–111, 2017. doi. 10.1109/CYCONUS.2017.8167504.

11 ISO/IEC 7498-1:1994. Information technology – open systems interconnection – basic reference model: the basic model, 1999. URL https://www.iso.org/standard/20269.html.

12 Jason Kick. Cyber exercise playbook. Technical report MP140714, MITRE Corporation, November 2014. https://www.mitre.org.

13 Lachlan MacKinnon, Liz Bacon, Diane Gan, Georgios Loukas, David Chadwick, and Dimitrios Frangiskatos. Chapter 20 - Cyber security countermeasures to combat cyber terrorism. In Babak Akhgar and Simeon Yates, editors, *Strategic intelligence management*, pages 234–257. Butterworth-Heinemann, 2013. ISBN 978-0-12-407191-9. doi: 10.1016/B978-0-12-407191-9.00020-X. URL https://www.sciencedirect.com/science/article/pii/B978012407191900020X.

14 James P McDermott. Attack net penetration testing. In *Proceedings of the 2000 Workshop on New Security Paradigms*, pages 15–21, 2001.

15 MITRE. A practical guide to adversary engagement. Technical report, MITRE Corporation, February 2022. URL https://engage.mitre.org.

16 MITRE. MitreATT&CK framework, 2023. URL https://attack.mitre.org.

17 Charles P Pfleeger, Shari L Pfleeger, and Mary F Theofanos. A methodology for penetration testing. *Computers & Security*, 8(7):613–620, 1989.

18 Hadar Polad, Rami Puzis, and Bracha Shapira. Attack graph obfuscation. In *International Conference on Cyber Security Cryptography and Machine Learning*, pages 269–287. Springer, 2017.

19 Chris Salter, O Sami Saydjari, Bruce Schneier, and Jim Wallner. Toward a secure system engineering methodology. In *Proceedings of the 1998 Workshop on New Security Paradigms*, pages 2–10, 1998.

20 Bruce Schneier. Attack trees. *Dr. Dobb's Journal*, 24(12):21–29, 1999.

21 Jonathon Schwartz and Hanna Kurniawati. Autonomous penetration testing using reinforcement learning. *arXiv preprint arXiv:1905.05965*, 2019.

22 Sugandh Shah and Babu M Mehtre. An overview of vulnerability assessment and penetration testing techniques. *Journal of Computer Virology and Hacking Techniques*, 11(1):27–49, 2015.

23 Nivedita Shinde and Priti Kulkarni. Cyber incident response and planning: a flexible approach. *Computer Fraud and Security*, 2021(1):14–19, Jan 2021. doi: 10.1016/s1361-3723(21)00009-9.

24 Yaroslav Stefinko, Andrian Piskozub, and Roman Banakh. Manual and automated penetration testing. Benefits and drawbacks. Modern tendency. In *2016 13th International Conference on Modern Problems of Radio*

Engineering, Telecommunications and Computer Science (TCSET), pages 488–491. IEEE, 2016.

25 Blake E Strom, Andy Applebaum, Doug P Miller, Kathryn C Nickels, Adam G Pennington, and Cody B Thomas. MITRE ATT&CK design and philosophy. Technical report MP180360R1, MITRE Corporation, March 2020.

26 Unal Tatar, Bilge Karabacak, and Adrian Gheorghe. An assessment model to improve national cyber security governance. In *11th International Conference on Cyber Warfare and Security: ICCWS2016*, page 312, 2016.

27 Clark Weissman. Penetration testing. *Information Security: An Integrated Collection of Essays*, 6:269–296, 1995.

28 Mehdi Yousefi, Nhamo Mtetwa, Yan Zhang, and Huaglory Tianfield. A reinforcement learning approach for attack graph analysis. In *2018 17th IEEE International Conference on Trust, Security and Privacy in Computing and Communications/12th IEEE International Conference on Big Data Science and Engineering (TrustCom/BigDataSE)*, pages 212–217. IEEE, 2018.

2

Overview of Penetration Testing

2.1 Penetration Testing

This chapter introduces penetration testing and an overview of what it involves. We will demonstrate modern cybersecurity practices at a high level and highlight where the processes are manually performed, slow, expensive, and less than ideal. After a quick look into some terms and historical activities, we will delve into the specifics and see how reinforcement learning (RL) can streamline the processes to aid the security professional better. Links to tools and example methodologies are provided to guide the reader to understand the concepts more quickly. By the end of this chapter, you should grasp red and blue team functions, how each plays a role in providing greater cybersecurity for their clients, and how the current process can be enhanced using strategic RL techniques.

2.1.1 Introduction to Red Teaming

In this section, we define the various teams that may be involved in penetration testing, briefly look into its history, and review some of the high-level concepts involved and desired objectives or outcomes of the teams' efforts.

2.1.1.1 Why? Reasons for Red Team Penetration Testing

Red team penetration testing is a strategic and operational cybersecurity practice used to test an organization's defenses and find its vulnerabilities thoroughly. Through simulating real-world attack scenarios and using the same sophisticated attack techniques as the malicious actors, red teaming exposes weaknesses that traditional security measures might miss. This proactive approach enables organizations to identify risks to critical assets, prioritize their limited security efforts, and fine-tune their defense mechanisms to respond to the latest threats.

Reinforcement Learning for Cyber Operations: Applications of Artificial Intelligence for Penetration Testing, First Edition. Abdul Rahman, Christopher Redino, Dhruv Nandakumar, Tyler Cody, Sachin Shetty, and Dan Radke.

Many corporations fall into strictly regulated data protection and privacy sectors, where high stakes and extremely sensitive information may fall into regulatory compliance mandates, and failing to maintain standards results in steep penalties and consequences. These regulations are in place to protect cyber infrastructure and the sensitive information contained within. Industry leaders and the government maintain and enforce most regulations. Ranging from the Payment Card Industry (PCI) standards [18], the Health Insurance Portability and Accountability Act (HIPPA) [8], and broad government cybersecurity standards [9], the regulations are continuously developed and updated to protect consumers and citizens.

Red teaming is integral in evaluating whether the company is meeting these standards. Ultimately, the red team penetration test empowers organizations to stay ahead of current and evolving threats, minimizing the impact of potential breaches and keeping their leadership informed. It may seem complicated, but maintaining cybersecurity is essential for continued safety and for the advancement of our digital lives. Some real-world benefits achieved by maintaining a policy of using well-executed red team simulations include:

2.1.1.1.1 Realistic Threat Simulation Red team penetration testing simulates real-world attack scenarios compared to static security assessments or in-place methodologies. These scenarios closely mimic the tactics, techniques, and procedures (TTPs) of actual threat actors targeting networks. Using red team tactics allows organizations to understand the vulnerabilities from an attacker's perspective and assess the level of risk to their assets and their ability to detect and respond to advanced threats.

2.1.1.1.2 Advanced Threat Detection Company networks contain and handle valuable and sensitive data, ranging from personally identifiable information (PII) to trade secrets, and are extremely attractive to cyber criminals and state actors. Red teaming helps identify the weaknesses that traditional security measures might not reveal. For example, vulnerabilities in networks, applications, appliances, cloud environments, and physical security. Detecting these vulnerabilities and remediating them before an attacker notices them secures the network from future breaches.

2.1.1.1.3 Evaluating Security Defenses Red teaming is not just a realistic test; it also evaluates the effectiveness of an organization's security controls. It assesses the capabilities of firewalls, intrusion detection systems (IDS), endpoint protection systems, and access controls. By determining whether the existing systems can withstand a sophisticated attack and provide timely alerts and notifications, the red team can provide an organization with a detailed report of the gaps and room for improvement in its existing security measures. The organization then takes these reports to its leadership, mainly the chief information security officer (CISO). The CISO weighs security needs against what they know regarding the company's risk profile to prioritize and implement the best cybersecurity strategy.

2.1.1.1.4 *Risk Assessment and Prioritization* It is repeatedly said that no network can be 100% secure. Understanding this fact allows red team penetration testing to assist in identifying high-risk areas that require immediate attention, especially if those areas contain high-value or crown jewel information. By exposing vulnerabilities and potential breach points, organizations can allocate their limited resources more effectively to mitigate the most critical threats.

2.1.1.1.5 *Compliance and Regulatory Requirements* Many industries are subject to strict legal and regulatory compliance standards (PCI, HIPPA, DDPR, SOX, etc.). Red teaming helps organizations demonstrate due diligence in security measure implementation and compliance with the overseer by identifying and addressing potential gaps in meeting regulations.

2.1.1.1.6 *Improving Incident Response* Red team tests frequently reveal how well an organization's security team responds to and implements its response plans during the simulated attacks. Documenting and comparing the results of a red team test to the blue team's security operations center (SOC) response can identify the gaps and areas needing improvement and collaborate in fine-tuning the incident response processes to minimize the potential damage and/or downtime during an actual breach event.

2.1.1.1.7 *Enhancing Security Awareness and Training* The weakest point in any network is the human. A successful red team penetration test, especially if it uses attack vectors involving human factors such as phishing, can be an eye-opener for the employees and stakeholders in corporations. It raises awareness about the sophistication of modern cyber threats and attack methodologies, as well as the importance of adhering to security policies and best practices.

2.1.1.1.8 *Executive Buy-in* Like the players going against the goalie in hockey, security defense teams must stop every attack, while the attackers only have to be successful once. While it's frustrating to be the defender, it can seem even more frustrating for the C-Suite that spends money on cyber defense measures and never "sees" the results. The outcomes of red team exercises can provide tangible evidence to senior management and executives about the potential impact of a successful cyberattack.

2.1.1.1.9 Third-party Risk Assessment Many companies and corporations rely on third-party vendors for various information-related services. Red teaming can assess the security of these vendors and virtual partnerships to ensure they are not introducing unacceptable vulnerabilities into the organization's digital ecosystem. These findings and situations commonly occur through business acquisitions and the desire for quick compilation of digital assets and minimal downtime.

2.1.1.1.10 Continuous Improvement of Security Posture Red teaming is not just a one-time event. It promotes a standardized continuous improvement cycle by allowing organizations to iterate constantly and refine their security posture based on the insights gleaned from each penetration exercise. Threat actors continuously evolve their TTPs and recycle and renew old exploits and vulnerabilities. The red team constantly trains and monitors these actors to stay consistently aware. Periodic red team tests ensure the latest security measures are in place and are effective against the newest threats.

2.1.1.1.11 Who – For Whom is the Testing? Understanding why the penetration test is essential creates a sense of its importance. Digging deeper into the personas that care about and benefit from the test gives a face to the reasoning behind these complicated security tests. In every organization that maintains cyber assets, most key stakeholders understand the importance of cybersecurity and the reasoning behind mitigating risks. However, increasing security is vital to every user and product on the network.

Among the key stakeholders, security teams, including the CISOs and Security Managers, are the primary focus and recipients of the testing reports. They gain strategic and valuable insights into the possible vulnerability and the effectiveness of security measures currently in place. Similarly, the IT department plays an important role. This department is responsible for the upkeep and running of the IT systems, so they need to comprehend how these systems are at risk of being compromised and how to respond to any incidents.

Executive management, including the CEO/CTO, also benefits by taking a keen interest in red team penetration tests. People at this level are concerned about overall risk management and safeguarding the company's reputation and assets. A red team penetration test offers a crystal clear understanding of the network security risks the business might face. The C-Suite teams' understanding of these risks is combined with their knowledge of business and management issues to create and implement a plan of action, even if it is as simple as approving budgets for the CISO's teams. Additionally, and as equally invested, are the compliance and risk management teams. These teams ensure the organization adheres to all regulatory and legal requirements for data protection and digital security. Meeting these regulations is often a core requirement for bidding, winning, and maintaining contracts. Demonstration of compliance also ensures that the company's reputation for cybersecurity is maintained.

A company's board of directors is another core group interested in the red team penetration testing results. Their vested interest is in guaranteeing that the company shields itself from cybersecurity risks that could have a meaningful financial and reputational impact. In a publicly traded company, the cybersecurity-based impact could severely damage the shareholder's interest or price for the stock.

The boots on the ground and operational teams running the business daily and handling the customer's business also benefit from red team penetration testing. These teams include salespeople, internal staff, and anyone involved with the business end of a company's interests. Ensuring the continuity of business and daily operations and protecting them from the impacts of cybersecurity incidents allows the work and livelihoods to continue.

Product and service development teams are another key benefactor of red team testing, especially in an organization that develops software or web-based/online services. Understanding how their products and services can be exploited is essential in maintaining a customer/client's trust in the organization's delivery. While the operations and product development teams may not be directly involved in the test, the outcomes and suggested remediations directly impact their work environment and data security.

The list of people affected by the red team's testing does not stop inside of the organization being tested. Auditors and external regulators frequently require evidence of adequate security practices and having the red team provide evidence of a successful penetration testing exercise can be a key part of this documentation. Customers and external partners, albeit indirectly, also have a vested interest in these red team tests. The security of their data, from financial transactions to trade secrets, relies heavily on the organization's cybersecurity posture and ability to defend itself from cyber threats.

Each group in the organization, from highly technical to daily operations and strategic to tactical, has individual perspectives on the importance of the red team's tests. The tests have a wide-reaching impact across many areas of a business's operations and planning and touch on many individuals working within and outside.

2.1.1.2 Teamwork: Red–Blue–Purple Teaming

The penetration test affects many areas of the business world and the people working within the organizations. In order to provide a test that is as effective as possible, the penetration testers often work in groups or teams.

Penetration testing, or red teaming, is a strategic test against a company's security measures. It is an *authorized* testing of a network's security systems, policies, and devices. The keyword here is authorized. Unauthorized testing of security systems is illegal and can be seen as a form of hacking to be hunted for and prosecuted. Penetration testing is typically considered a proactive security measure, as opposed to reactive, because the end product will result in recommendations for implementation that increase the test target's cybersecurity posture.

Cybersecurity teams, especially penetration testing teams, can go by many names in today's world. They are pen testers, ethical hackers, and white/gray/black hat hackers. The teams cybersecurity work on while working in penetration testing roles are usually known by various names such as red, blue, or purple "teams." Whatever they are called, every security team can have personnel specializing in highly technical areas of cybersecurity or can be a blend of security experts and expert researchers. A quick description of the various types of teams observed in today's security environments gives us a high-level understanding of what they specialize in. Remember that teams may work independently of each other or collaborate in a more extensive security assessment.

2.1.1.2.1 Red Teams Red teams are made up of offensive security experts that are trained and experienced in thinking like the enemy (an attacking force). They have tools and training that allow them to simulate attacks, perform detailed scanning of network infrastructure (internal and external), and implant computers with simulated malware.

2.1.1.2.2 Blue Teams Blue teams are made up of defensive network security experts. They are trained in ways to identify network attackers, recognize malicious behavior, and have an understanding of the internal network defenses and how/what should be going on in a network, their job is to ensure all security measures are in place and functioning properly.

2.1.1.2.3 Purple Teams The purple team is not necessarily a separate team engaged in an active security test but a collaborative effort where the blue and red teams work closely together. Purple teams are the experts who know both red and blue team members' skillsets. The purpose of the purple team is to assess the security of a network from an offensive and defensive standpoint, using the experiences of the red team to drive effective translation of defensive strategy improvements. These recommendations are based on the blue team's expertise in forming and implementing new network security measures.

2.1.1.2.4 Threat-hunting Specific Teams Threat-hunting teams are usually called in after an attack or the suspicion that an attack or infection may have occurred. While the details of threat-hunting operations are complicated, many of the same skills and knowledge overlapped between them and the red/blue/purple teams.

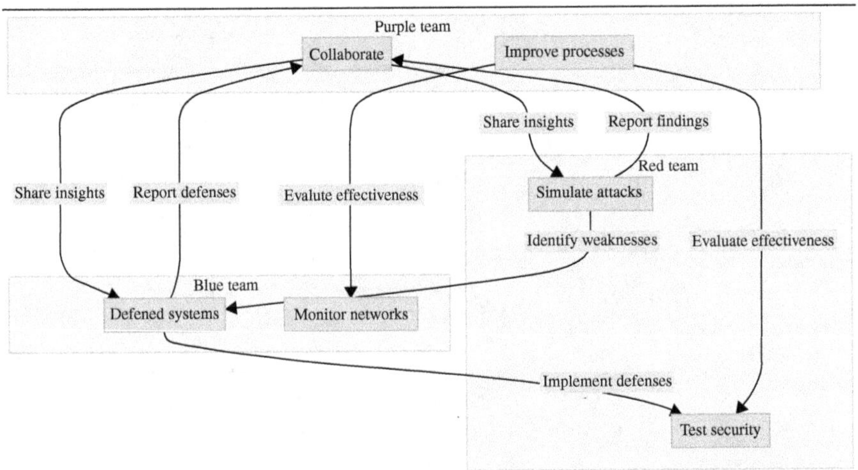

Figure 2.1 Red, blue, and purple teams working together to secure the network.

Threat-hunting teams hunt through a network, pulling in data and forensic information to discover what happened or could have occurred when an intrusion is suspected. It is common for the responsibilities and roles of the various teams to overlap. Whether the teams know about each other, working independently or as a part of a coordinated effort, the end goals are the same. Find vulnerabilities and potential network risks and report the findings logically to leadership (Figure 2.1).

2.1.2 A Brief History of Red Teams What and Where

Penetration testing, specifically red teaming, became a common practice as far back as the internet was invented. Militaries have used red team exercises for years to simulate enemy combat situations and test ideas and countermeasures for as long as conflict has existed. Straight from the US Army's "The Red Team Handbook," Cyber red teams operate primarily with "Outside-In Thinking" [2]. Seeing as the military developed the first backbones of the internet, it only makes sense that once security was finally thought about, required, and implemented, ways to test it were invented in parallel.

Security was not precisely a forethought of the internet inventors' minds. They were inventing something new and high-tech. The early internet, shared exclusively among researchers, colleges, and the military, was small and trustworthy. Security quickly became a necessity once the internet got into the hands of the public. It is even more prevalent and relevant today, where the number of devices and possible connections is growing exponentially.

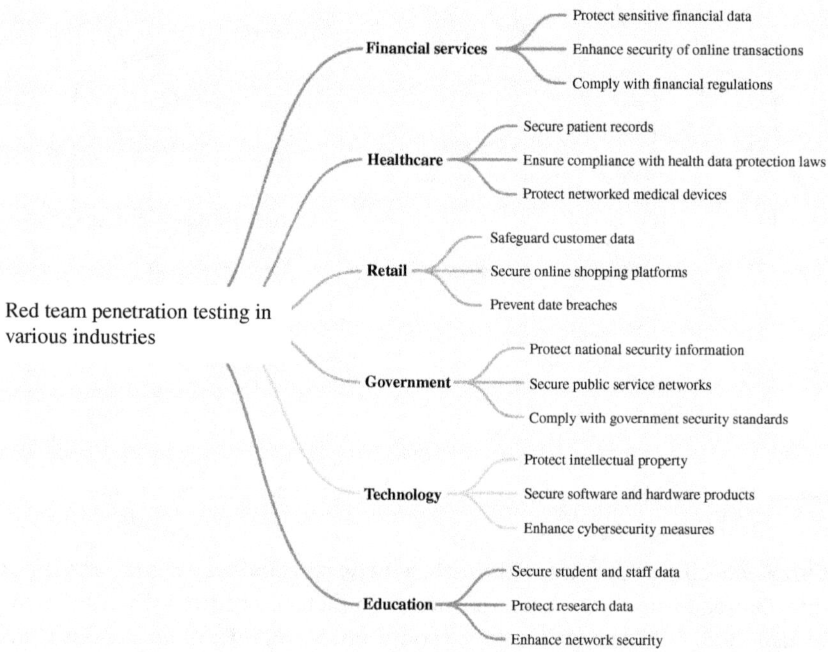

Figure 2.2 Red, blue, and purple team in various industries.

Penetration testing may be a slow and manual process, but the benefits have been outweighing the costs thus far. It is used by practically every industry that uses computers for operations, data storage, or financial transactions. Digitization of records, particularly in the healthcare industry, and the drive of every sector to be competitive has put many services previously reserved for in-house use to cloud platforms. People want to be able to access, move, trade, and buy stuff with their money in a convenient, secure manner. Below, we will look through some of the most prominent industries affected by cyber and discuss how red team penetration testing helps the companies and industries (Figure 2.2):

As seen in the chart above, pen-testing and red teams affect many aspects of business and personal matters. Let's examine some specific ways red teams and industry sectors work together. Look at all the areas that red teams protect and secure.

2.1.2.1 Military, Government, and Defense Industry

The United States Department of Defense (DoD) realized that cyberspace was a new domain of combat and operations that necessitated "comprehensive computer security and defense" [1]. Military programs to secure and defend computer networks existed as early as 1972, previously grouped under various intelligence organizations. Still, in 2010, USCYBERCOM was established due to the increased threat to public safety observed by rising levels of hacking, cyberespionage, equipment malfunctions, and the world's reliance on cyberspace resources. The protection of military assets demanded that protocols and procedures be put in place, and what better way to test a system's vulnerabilities than using hackers and their curious nature? In fact, the DoD found the exercises so valuable that they hold a joint-force exercise every year, where each military branch competes and strategizes in a series of war games. They perform network operations in a series of exercises with a military red team simulating enemy activity [10].

2.1.2.2 Financial Services and Commerce

Financial industry sectors, like banks, insurance companies, payment card corporations, and other financial institutions, rely on penetration testing to protect complex and highly targeted cyberinfrastructure. Money makes the world go round, and criminals trying to steal that money make the entire industry a target for attack. The requirements for protection extend beyond the protection of customer PII data to maintain trust; it encompasses a wide array of security challenges unique to this industry.

First and foremost, the online banking portals. These websites and their infrastructure are the portal customers and businesses use to manage their financial assets and interact with the banks. These platforms are a historically lucrative target primarily due to the amount of personal data and financial information they can reveal. Red teams must know about secure technologies such as web applications, mobile web apps, backend database systems, and encryption technologies.

For money to be valid, it must be transferable, or in other words, spent and received. The PCI has evolved into one of the largest industries dealing with this funds transfer between buyer and seller. Payment cards are used at ATMs to withdraw money and at every point-of-sale (POS) terminal. When you tap, swipe, or dip your payment card into the POS terminal to pay for gas, there is an entire network of digital communications behind the scenes verifying the card's legitimacy, funds available, and other transactional information. All these communications must be secured from interception, decryption, or fraudulent activities. Red team penetration testing once again identifies the potential security weaknesses in these systems that could be exploited for unauthorized transactions or data theft.

Investment banks and platforms within the financial sector also rely on moving money around. A vital component of the stock market and other investing platforms that trade digitally is the trust and speed of its networks. Success in this arena, where milliseconds can mean millions in loss or profit, is pivotal to high-speed secure networks. Red teams simulate attacks and test security against these systems to ensure reliability, safety, and resilience to prevent disruptions in trading activity or manipulation of data in the markets.

A fundamental way to ensure companies comply with regulations and maintain customer trust is to have an accepted set of standards and principles. The rules this industry has adopted are the payment card industry – data security standards (PCI-DSS) [4]. These standards verify that any system used to handle credit card or financial data is secure and up to an acceptable level of security.

Every business in today's society uses the internet in some way, shape, or form. Whether it's your local business or plumber that uses online technology to handle schedules, advertising, service selling (product selling, or the Fortune 100 company with massive storage and processing networks, they are all vulnerable and must be tested. Data has effectively become the world's most valuable resource. A company's private data is even more helpful. Companies have to collect and protect (by law) sensitive personal information (PII), guard company tradecraft and proprietary information, and ensure their employees have the access needed to use the information when required.

By simulating realistic cyberattacks and observing vulnerable systems, red teams can protect PII and ensure systems meet or exceed industry regulatory standards. Nobody wants to connect to the bank's website to check their balance and worry that their information will be observed by malicious actors. Red teams can test these connections for weaknesses.

2.1.2.3 Healthcare

Red team penetration testing is critical in safeguarding the digitally interconnected healthcare ecosystem. From patient records stored in the electronic health record (EHR) systems to billing, the only part of the industry that doesn't involve cyber is the doctors and nurses. The Department of Health and Human Services and its Office for Civil Rights (OCR) enforce stringent laws and regulations around patient privacy, known as the HIPPA [6]. Failure to maintain these cybersecurity standards results in stiff penalties, up to $250,000, and 10 years in prison, along with damage to the company's reputation and ability to operate in the future.

To maintain security standards and avoid violations, the red team exercises rigorous tests against the security of any EHR system. Any system that contains patient data must be compliant. Identifying and mitigating vulnerabilities and bad security practices in these systems protects patient information from unauthorized disclosure or access and ensures compliance with the laws. Beyond patient records, red team testing is crucial for ensuring communication stability within and without the healthcare facilities. Telecare and virtual consultations are becoming regular parts of the industry's affordable healthcare options, and ensuring these communications are secure enough to maintain doctor–patient confidentiality is vital in maintaining patient trust.

Lastly, stepping beyond the healthcare infrastructure, medical devices are becoming more complex and cyber-enabled. Instruments like pacemakers, infusion pumps, and diagnostic equipment are regularly connected to the network for updates, data transmission, and diagnostics. Red team testing and research assist in uncovering exploits or vulnerabilities in these devices and their communications, which could be used to harm humans or exfiltrate sensitive data.

2.1.2.4 Technology, Telecommunications, and Cyber

In the fast-paced, innovation-driven, and intertwined worlds of the technology and telecommunications industries, red team penetration testing is not just an afterthought security measure but a cornerstone of cyber resilience and trustworthiness. Technology and telecommunications rely on software development platforms, cloud and local computing resources, networking technologies, and data storage solutions to create, maintain, and deploy technical solutions to their customers. Any mistrust or loss of capability in these systems costs the companies not only money or losing proprietary code but also the consumer's trust and the company's good reputation. This is why companies rely on red teams to ensure the many facets of their cyber landscape are protected.

Software products are developed on technology systems. Red teams assist in the software development lifecycle (SDLC) from the early stages to post-deployment. Software development can be further secured with the assistance of red teams by simulating real-world attacks using automated code-checking and manual pentesting techniques [38].

Red team penetration testing is also pivotal in safeguarding the cloud from attacks. The penetration testers ensure the safety of data storage, processing, and computing, access management systems that run on multiple cloud vendors, and locally hosted cloud technologies. Cloud systems can store sensitive customer data and run critical business systems such as commercial retail applications. Securing cloud-based systems is as important as hardware systems since they are remotely hosted and potentially even more vulnerable.

Within the world of telecommunications, the red team is vital for protecting the infrastructure that is the backbone of global communications. Plain old telephone (POTS) telephone networks have mostly given way to Packet Switch networks that run on extensive transmission control protocol/IP (TCP/IP) networks, aka the internet. Red teams test network equipment and mobile phone networks (including wireless protocols) while ensuring the security protocols provide a robust defense against eavesdropping, service disruptions, or data interception.

Another critical area to consider in technology is the Internet of Things (IoT) devices. IoT devices have become increasingly integrated and integral parts of the tech industry [37]. Millions of new devices are being internet-enabled every year.

Devices from home appliances, like dishwashers and clothes dryers, to industrial control systems/supervisory control and data acquisitions (ICS-SCADA) systems collect and send data over the internet. These devices can control critical public infrastructure, from water to electricity, and any successful attack would harm the public.

2.1.2.5 Conclusion

After observing these areas above, it should be obvious how vast the world of cybersecurity is and how many places it can touch in our lives. Protecting these technologies and assets is a global priority. Maintaining the battle rhythm and consistent demand for all the industry sectors above is a cumbersome requirement. One that is not going away any time soon or getting easier to do correctly.

Unfortunately, modern penetration testing is structured, regulated, and often slow or cumbersome. This is due to the sheer volume of information, the required knowledge, and the research to be performed. In further chapters, we will discuss applying artificial intelligence (AI) machine learning theories and technologies to assist the red team in achieving this difficult but necessary goal.

First, let us look at the current concepts, methods, and data requirements the modern red team deals with daily

2.1.3 Modern Penetration Testing

Modern penetration testing requires many complex steps, testing many potential theories against possible weak points on multiple systems levels and possibly on different architectures and networks. These complexities lead to a slow and manual process, even using frameworks and modern toolsets. This chapter will cover the basic concepts of penetration testing, iterate the most common steps and objectives, and introduce some of the methodologies along with well-known and commonly used frameworks that exist. Along the way, we will pause to examine the human role in investigating machines and think about how we can use machine learning to assist the pentester and make their job faster, easier, and more cost-effective. In this section, we will discuss the main concepts surrounding a penetration test. Topics include overarching concepts to familiarize the readers with how red teams must prepare for, execute, and report on their work. It is essential to remember these subjects while trying to understand how machine learning can assist in automating, speeding up, and improving them. We will look at areas such as the types and styles of engagements (black/white/gray), the objectives of the test, and the methodologies involved.

2.1.3.1 Types and Styles of Pentesting Engagements

Penetration testing teams can start with different levels of initial knowledge when it comes to the desired depth and required work/research. From an organization's point of view, the three industry-accepted styles of penetration testing are black box, gray box, and white box. Each style has its own level of starting knowledge and different amounts of time and resources required to perform. Of course, the trade-off for faster, less expensive testing is the possibility of less in-depth reporting or findings. More resource-intensive testing may seem like the best solution. However, situations exist where the penetration testing team must trade in-depth research and analysis for less detailed answers in less time. Each test must begin with a conversation amongst stakeholders that discusses the pros and cons of each level of testing. This way, the penetration testers, and the end customer understand the goals, limitations, and scope before performing any actions.

A summary of some of the advantages and disadvantages of penetration test types is found below in Table 2.1.

The choice of methodologies depends entirely on the company and the desired goals of the testing. Factors that come into play that can drive the selection include the amount of time available, desired engagement accuracy and depth, and level of coverage required (legal, etc.). Money/budget for the security teams also comes into the conversation. Full-depth investigations may take longer and cost more money, along with time. Management and IT must weigh the tradeoffs between the accuracy of the tests and the achieved level of coverage against the speed, efficiency, and depth of coverage.

2.1.3.2 Black Box

Black-box testing is the most involved and considered the most challenging level of penetration testing. In this level of testing, the penetration testing team is given the bare minimum amount of information. From there, it must gather the knowledge needed for a successful test. The data may be as sparse as the company's or target's name, an IP Address, or a URL.

The team performing a black-box test must utilize their research and investigative skills to dig through their information sources. They take the gleaned information and pivot as much as possible to get the most accurate picture of a company's cyber footprint, also called its cyber terrain. The team will take this information and feed it into the methodologies discussed (previously) for use in the full-on test.

This type of test can be the fastest one performed. The price of speed, however, is that some details or vulnerabilities may be missed along the way. However, the black-box test can also be a starting point. The information uncovered during a black-box test can be shown to the company stakeholders, potentially reasoning for an escalation to deeper or broader testing techniques.

Black-box testing is used when a company desires a full-on test that accurately mimics the situation of an outsider. This test may be slower and yield fewer results, but it provides a picture of the company's cyber terrain. Due to the limited amount of information given to the team from the start, blackbox testing requires a deep

Table 2.1 Levels of testing compared to speed accuracy and depth.

Pentesting level	Level of initial knowledge	Pros	Cons	Depth of test coverage	When to use
Black	Almost none, minimal	Faster, mimics real world end-user experience	Less details more probability of missed vulnerabilities overestimation of results	Least potential discovery	When time is short, or a test of external points is all that is desired Budget friendly or tight deadlines Need real-world realism.
Grey	Some starting points, minimal internal network information	Deeper then black, not as time intensive as white (balance) communication between admins and testers is encouraged	More time intensive then black-box slower-speed sacrifice for efficiency and coverage	Medium discovery traded for reduction in time	When there is more time and a deeper level of information is desired Desired collaboration with developers/admins More complex systems are in use compliance
White	Full access and deep starting knowledge	Most comprehensive Allows indepth target system knowledge, more potential discovery, ensures close collaboration between admins and penetration team	Takes the most time and resources of all three. More costly due to expertise in expansive areas and hours.	Full discovery possible, access to full network, logs, and resources.	A full network assessment is desired, time is not critical Critical that dev/admin be part of the collaboration Compliance standards or regulations demand it Testing against insider threats.

understanding of network technologies, research methodologies, and investigative tool usage.

2.1.3.3 Gray Box

Gray-box testing is the next step up from black box. The penetration testing team is provided with more information to get started in the gray-box test. This could be information surrounding specific hosts or target networks or information involving technical details that might not be public knowledge. Lists of known IPs that belong to the company, email domains, URLs that the company owns, and types of infrastructure the company assumes it owns are among the usual starting points. Everything else must be discovered through testing and research.

This type of test provides a good idea of what a targeted attack might be without the penetration testing team spending time performing the initial investigations and developing the attack footprint. Further research can be conducted to ensure that the provided information is accurate and uncover data that may have been missed or unknown to the original providers of the source information. One example could be test networks or development environments in the same ranges uncovered during the normal reconnaissance phase. It isn't uncommon for production managers to be unaware of a developer team's testing networks, and sadly, it is even more common for security not to be a priority in development environments.

Gray box testing is a valuable way to reduce the time required to start a full penetration test. Companies may still want to get the whole experience of seeing how their network looks to an external hacker but may opt to give the red team a decent baseline of information to speed the initial research and investigation phase along. Gray box testing may also be used when complex systems are in play or little existing documentation exists. The system administrators have not kept the network documentation up to date, or the system's footprint may have proliferated so that no one internally knows 100% what exactly is there. It's also possible that the company/customer only desires the red team to focus the test on specific parts of the infrastructure. In such situations, the red team can take what knowledge objects exist and go through the concepts discussed (earlier) to pivot and expand their attack surface.

2.1.3.4 White Box

White-box testing, also called glass box or clear box, is the most in-depth and comprehensive penetration testing. This testing focuses on both external and internal vulnerabilities. Access to the internal networks is usually given to the penetration testers from the get-go. The penetration testing team is provided with internal documentation, configuration schematics, initial scan and known endpoint data, and potentially historical log data. With all this data from the get-go, machine learning/RL techniques can go a long way in assisting the red team. Log analysis, network best-path generation, and other strategies will be discussed later. Think of all the opportunities!

Benefits to this type of testing include the team spending more time finding the vulnerabilities and performing more targeted network scans. The information is provided upfront, and the team doesn't have to do as much research. Without initial access scanning and internal network mapping, the team can potentially uncover previously unknown or undisclosed network sections or endpoints and their vulnerabilities.

White-box testing is also usually performed hand-in-hand with the system administrators and developers, which can supply them with in-depth information about the network and endpoints when requested. This information exchange allows for clarification of details, avoiding misunderstanding, and spending unnecessary time digging around targets of low interest or low value to the company. There isn't much use digging around a testing or benign network of endpoints when the crown jewels are entirely located in a different place.

Some situations will require a white box test. For example, when evaluating a new technology still in early-stage development. In this situation, it is vital to incorporate white-box testing into the coding and development stages to ensure security vulnerabilities are discovered and patched before they are released into the wild. Another situation that warrants a white box test is the identification of code-level errors. Identifying coding errors, weak algorithms, or insecure coding practices requires the code to be revealed and assessed along the way. Finally, white box testing is a requirement for many stringent regulatory compliance requirements. Usually, the regulations stipulate that an exhaustive examination is performed on all aspects of the code and (network) systems used to prove that an organization performed all due diligence in assuring every system component is free from obvious flaws and vulnerabilities.

White box-type testing is usually the slowest and most expensive but results in the most in-depth coverage and assessments due to its open nature and inspection of all aspects of the environment. This isn't to say it's without challenge, but many of these slower processes may be sped up with the help of AI RL technologies discussed later in the book (Table 2.1).

2.1.4 Objectives, Considerations, and Goals During a Penetration Test

As a proactive and essential part of any cybersecurity strategy, the penetration test is usually performed with objectives or goals in mind. Any security test aims to always locate and secure any existing vulnerabilities. Along with the topics and industries covered earlier in this chapter, it is essential to consider other potential use cases and objectives of performing a security test.

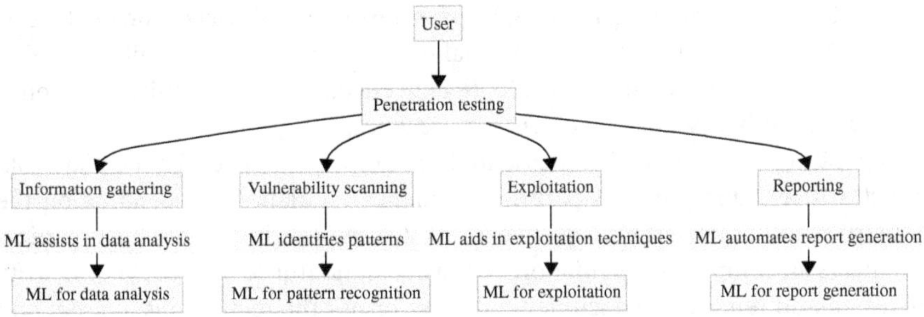

Figure 2.3 Machine-based penetration testing assistance workflow.

The penetration tester must account for many factors while going through the steps and methodologies discussed later in this chapter. Variables include network locations, the current level of operating privilege, the level of monitoring and logging, and the desired outcome of the current step. Remember, during the penetration test, the red team members always try to think like an adversary, often taking the same steps while looking for vulnerabilities. While the end objective of the adversary usually has some malicious intent, the red team will perform the same actions to find weaknesses and vulnerabilities to fix and report.

The reader should take note as we progress through the testing objectives and considerations about where machine learning and other AI technologies can assist with the red team's final goal: finding vulnerabilities and remediating the risks. Any task that can be sped up, made into an automated workflow or requires digging through data can benefit from machine-based assistance.

The chart below shows an example workflow where machine-based assistance helps an operator by automating several steps or organizing/filtering/analyzing the data to highlight important aspects. We discuss ideas at the end of each phase, however, technology is evolving at a rapid pace and the possibilities for ML implementation continues to grow (Figure 2.3).

2.1.4.1 Objectives

The objectives of a red team test vary from engagement to engagement, but the core responsibilities and goals will remain roughly the same. The primary objective is always to uncover security weaknesses in systems, networks, applications, and appliances before the attackers do. Each goal's details are all to act and think like an enemy would beat them to the vulnerabilities. To evaluate the environment's security, it is occasionally necessary to assess the effectiveness of existing security policies and protocols to understand how they can withstand an actual cyberattack. The policies can be evaluated by reviewing the policy manuals and documentation or by actual world application of red team techniques. Testing the policies in real-world situations may result in the blue team or network SOC getting alerts that something is amiss. Depending on the test's level and depth, this may be the perfect time to test the internal incident response capabilities. Testing the response capabilities aims to evaluate how effectively the organization identifies and responds to security incidents or breaches.

Responding to breaches is an essential part of any security team's playbook. In fact, it may be required as part of compliance regulations, depending on the industry. Another important objective for red teams is verifying that the organization meets all compliance regulations. Industry standards related to cybersecurity are tested and reported to document whether or not the regulations are met and how to fix them. Red team activities also maintain the constant objective of assisting with risk assessment. The test itself is performed by looking for vulnerabilities and weaknesses. The reporting and reviewing after the test helps an organization's leadership assess the amount of risk the network has and how it relates to business operations. Penetration testing allows leaders to weigh the risk against the costs of an incident and recommends steps to take to reduce that risk. One of the most significant risks, and thus objectives for the red team, includes data breach protection. The red-team's test objective in this regard is safeguarding sensitive data. The data is protected by identifying and mitigating the potential breach points and ensuring the sensitive information is adequately protected and cannot be easily accessed or stolen.

2.1.4.2 Considerations

Having a set of goals and objectives is an integral part of the penetration test, but there are some considerations the red team must keep in mind while performing the test. Keeping the information below in mind will ensure the red team does the test effectively while maintaining system integrity and staying within legal frameworks. The considerations discussed below are an essential part of successful tests. Making the wrong choice during a test could devastate the company's network, the red team's successful execution, or the legality of any regulation requirements documentation [39].

Red teams must always consider what they are allowed to do. What is out of bounds, and what is in play for the testers? This consideration is known as the Scope and Boundaries of the test. Clearly defining and outlining the scope of a penetration test informs the participants on what will be done, what systems it will be done to, and when the test is considered successful or complete. These considerations can list which systems, networks, and applications are to be tested as well as the goals of the tests.

Having the goals of the test line up brings the consideration for legal and ethical compliance. Ensuring that the penetration test complies with all relevant laws and regulations to avoid legal repercussions prevents the team from getting into situations they may wish to avoid. Considerations surrounding the legal and regulatory issues should be discussed and outlined with the company's stakeholders before starting any engagement. This ensures that all testing activities are done with the company's knowledge, are entirely lawful and ethical, and all necessary permissions have been obtained. A big part of the ethical consideration comes down to ensuring the test is conducted ethically and doesn't cause undue harm or disrupt the organization's operations or profit-generating activities.

Having ensured all the legalities and ethical considerations are in order, the red team must consider which testing methodology to use (black, white, and gray box). This will depend on variables and resource constraints discussed with the company's stakeholders during the planning phase. It is essential to consider all the team's objectives and resources before jumping into the testing procedures. As discussed in this chapter, testing methodologies will determine many other considerations and testing variables (time/length, resources required, skills, etc.) [42].

As well as figuring out what and how to do it, the red team needs to consider the communication channels used during the engagement. Effective communication between the team members and leaders during the engagement ensures smooth operation with little overlap. It is also essential to maintain effective contact with stakeholders. These communications need to establish methods and protocols for reporting findings and incidents that may occur and keep the stakeholders informed about the tests' progress.

Even the most careful planning and communications considerations may not prevent incidental outages or impact on network bandwidth. Considering the timing and duration of the network penetration test will prevent any disturbances. Teams should decide on an appropriate time and duration for the test to minimize the impact on regular operations while still getting accurate pictures of the network security posture.

Now that the test is about to begin or is underway, the team has to carefully consider the tools and techniques. They must utilize the appropriate tools and techniques to simulate a real-world attack scenario to the best of their ability. This may be done by selecting known tools and data sources like the ones discussed in this chapter or using custom-written scripts and tools. The environment where threat actors operate is vast and practically unlimited in scope, so choosing the right tool isn't as much of a concern as being familiar with the tool. Questions like how the tool works, what data it can return, what scans it can do, and how effectively it can be used for this particular engagement should be considered before selecting and using the toolsets.

Once the tools are being deployed and the engagement is proceeding full steam ahead, a lot of data will flow in. Data is the rock and debris the team sifts through to assess and find vulnerabilities. The crew should consider how they will handle that data to succeed in the goals laid before them. Ideas should be in place for managing the data volume, data storage, network capacity, and what tools will be used to analyze the final collection(s) of data. Also, as necessary, the team must consider how situations will be handled if sensitive data is encountered. The planning stage should set boundaries on what data is off-limits and what to do if that type of data is located. Examples include live operational data, PII, and credit card numbers. If collected and in scope, this type of data needs to be collected and handled securely during the test to prevent data leaks, breaches, or unethical activities.

Phew, the test was extraordinary. Now, the team has to consider what sorts of follow-up actions will occur. Planning for remediation of the vulnerabilities and retesting is needed to confirm that the security issues have been resolved and the vulnerabilities mitigated. These results will be fed back into post-test analysis for future work. The post-test analysis considerations will also provide actionable insights and recommendations for security improvements to the company stakeholders. These results will (hopefully) demonstrate to the company that regular testing and updates should be an ongoing process, not a one-time event. Findings should drive improvements, which should be tested again later.

One final and often overlooked requirement is documentation and reporting needs. The object of the test is to report the findings to the customer so they can be remediated, thus improving security. Considerations and plans must be in place to ensure that careful notes and documentation on testing methodologies, findings, and reported recommendations are being made. This documentation helps the red team show its work and align with all the other goals and considerations should a situation arise. It is also a good practice to keep track of what happened so that the team members responsible for performing analysis know what steps led to which data being found and which vulnerabilities were discovered.

The prior sections provided some high-level details about the red team penetration test and what it looks like from a planning and consideration point of view. It should be generally understood why the type of team that was selected was put in place and what their objectives and considerations are. Now, we will dive into some of the methodology and tools red teams use while performing a client engagement, also known as the penetration test!

2.1.5 Methodology

Methodologies are systems or sets of rules used by any professional discipline to carry out its work or generally carry out research [12, 15]. Regarding red team penetration testing, a methodology refers to the organized and structured procedures performed during a penetration test. A methodology might be a combination of processes and guidelines for conducting penetration. It helps the testers narrow their focus, follow specific steps to competition without failure, and ensure the test retains its validity in the industry [7]. In this section, we will discuss the

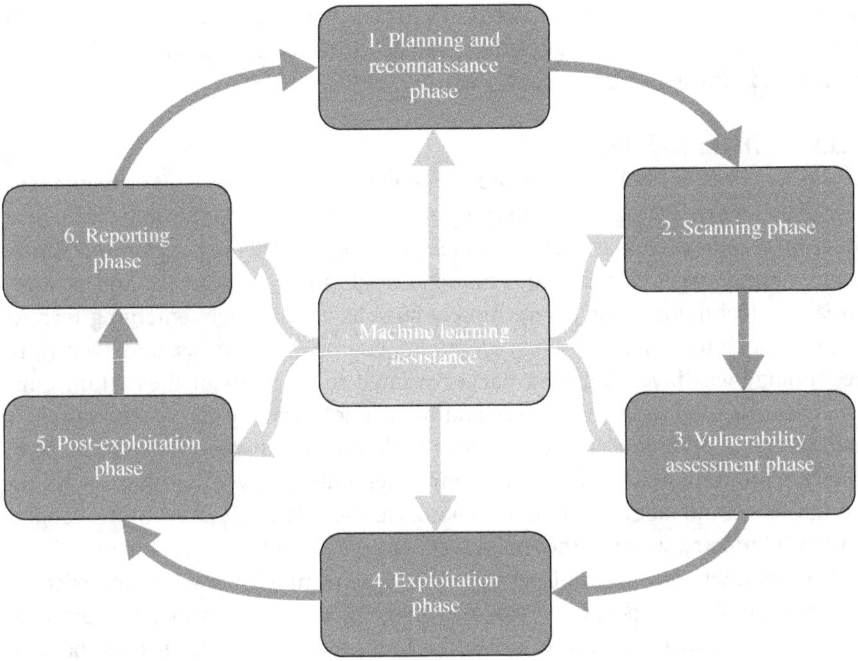

Figure 2.4 Pentesting lifecycle and where machine learning can assist.

required mindset, go over the phases of a penetration test at a high level, and end with some considerations for reporting and disclosures. Along the way, the reader should consider how the phases will lead to data that can be utilized for machine learning and RL purposes. The end goal of the test is to find vulnerabilities and mitigate them. Still, the intent of this book is to guide the reader into thinking critically about red teaming and how the processes can be made faster, more efficient, and more enlightening than the current practices are (Figure 2.4).

2.1.5.1 Thinking Like an Adversary

To perform pen-testing properly, the practitioner must adopt the mindset of a hacker. The word hacker has a negative connotation in modern-day news and tends to be in newspaper headlines whenever and wherever something terrible happens in cybersecurity. The truth is that most red team members have to be pretty decent hackers to be the best at what they do. Hackers, or hacking things, were not always considered bad things. As we will see, it originally referred to a curious individual with a spirit of adventure and the mindset to use tools for purposes other than the intended designed ones. Modern-day hackers are experts not only in networking sciences but also in experimenting with all kinds of technology. AI, machine learning, and RL play right into the data science/hacker mindset, with potential uses limited only by imagination and the willingness to explore. Bad guys think outside the box and use any available tool or method. The good guys must as well.

2.1.5.2 The Hacker Mindset

A hacker was initially considered someone with an insatiable appetite and curiosity for everything related to computers, technology, and the digital universe. Before "life hacks" and "10 ways to hack your foo bar," some people were simply overwhelmed with curiosity and found solitude in tinkering with and possibly breaking technology. They were merely curious about newly emerging technological capabilities surrounding computers, computer networks, and everything technology was inventing. The hacker wanted to learn about these things and then take that learning and understanding to a new level by deconstructing them and using them in new and novel ways. Technological discoveries were made by hackers building new hardware in their basements or inventing novel ways the computers could talk to each other. Thanks to the original hackers and leaders of the tech industry, we have the modern cyber ecosystem that exists today.

However, wherever there is an excellent opportunity for good, there exists an opportunity for bad people to take advantage of the system as well. From tried and true con-artist type scams and script kiddies defacing websites just because they can, ransomware gangs out to make a quick buck, and up to nation-state advanced persistent threats (APT), these modern-day criminals (people performing criminally defined behavior), were labeled as "hackers" and the term stuck. An infamous hacker once said, "Yes, I am a criminal. My crime is that of curiosity." [40].

To be an effective red teamer, one must understand the technology used, the mindset and intentions of the malicious actors, and what they desire to accomplish. From attack surface mapping to optimizing exfiltration routes, network mapping to vulnerability enumeration and research, the red team explores every possible vulnerability to ultimately report the findings to leadership and the cyber defenders, and finally plug the holes and prevent malicious actors from achieving their goals. Technical prowess and ingenuity go a long way in red teaming, but these gifts don't change the fact that art is complicated. It also takes time and resources and can be highly inefficient. With the constant struggle to be on top, taking advantage of every tool and technology you can is the only way to stay on top. This is where machine learning can give the red teamer a leg up while doing their job, making the entire process faster and more efficient. AI technologies won't lead to fully automated solutions or the machine doing the work for you, but they can take the data gathered and analyze it more effectively and efficiently. This results in less manual labor, digging through tons of data, and more time using the human hacker spirit to dig into all the potential holes, pivots, nooks and crannies, and network infrastructure. It also helps guide research in the right direction for the right jobs and helps summarize the findings for presentation to policymakers and action-takers. Modern penetration testers follow a structured framework to ensure the process is predictable and documentable. Standard-based methodologies and frameworks have been created and tested over the years. The pen-testing community has accepted many as best practices and are available for use. The pen testing team lead will be familiar with various frameworks and how they apply to each engagement to accommodate the customer's situation. They will choose the best one based on their experience and the test requirements. Several industry standard frameworks are used often and occasionally interchangeably, so we will simply mention them in no particular order and cover them at a high level before going into specifics about the methodologies.

A summary of several industry-accepted penetration testing frameworks is listed below:

1. **Open Source Security Testing Methodology Manual (OSSTM)**: This framework is a peer-reviewed security testing methodology maintained by the Institute for Security and Open Methodologies (ISECOM) [5]. The OSSTMM provides guidance on testing operation security by breaking it into five concepts: physical, wireless, telecommunications, data networks, and human safety. It offers a scientific process to accurately measure and characterize security in understandable terms.
2. **Open Web Application Security Project (OWASP)**: OWASP is a nonprofit organization that desires software security. The community of open-source developers maintains guides on security testing, including Web, Mobile, Network, and Firmware Testing Security Guides. They also maintain the OWASP Top 10 list of vulnerabilities and focus on ways of securing and eliminating vulnerabilities on the list [17].
3. **National Institute of Standards and Technologies (NIST)**: NIST provides a set of guidelines and best practices for penetration testing, outlined in various publications like NIST SP 800-115 [13]. The US Government agency provides multiple standards and frameworks related to cybersecurity and the security controls listed by NIST. Organizations can be held to these standards to win specific government work.
4. **Penetration Testing Execution Standard (PTES)**: PTES is a comprehensive guide well-established in the field and maintained regularly. It defines seven phases of penetration testing and hands-on technical guidelines for what and how the test should go, the rationale behind the testing, and recommended testing tools and how to use them [3].

The various methodologies and existing frameworks have been summarized in an easy-to-understand section below. The list of potential methodologies may be outlined, but the discussed information touches on all significant aspects of penetration testing, primarily where they apply to RL and AI technologies. Should the reader desire to dig deeper into the methodologies or standard organizations, links will be provided in the references/citations section.

2.1.5.3 Pentesting Phases

Pentesters have continuously evolved and honed their techniques to improve speed and efficiency. While there are many improvements to be made, especially with the added assistance of machine learning innovations, the basic steps have been narrowed down to cover six distinct phases. Other standards may have more or fewer phases and go into various levels of detail. Still, the phases below provide a universal starting point for understanding what happens during the overall penetration test. The phases essentially provide a way to group pentesting activities into an easy-to-understand and logical order.

- Planning and Reconnaissance Phase
  - This phase involves gathering as much data and information about the target system(s) as possible, including network topology, OS, users, and applications.
- Scanning Phase
  - Scanning involves using tools to identify as many potential points of interest or points of entry as possible. Scanning for open ports, running services, network traffic patterns, etc. (e.g., Nmap, Wireshark)
- The vulnerability assessment phase involves using data about points of entry/interest and information gathered during reconnaissance to identify potential vulnerabilities and ways to exploit them. Research and database checks are performed (NVD, common vulnerabilities exposers [CVE], etc.)
- Exploitation Phase
  - All data gathered and research performed, this phase involves actively testing and attempting to access the target systems, exploiting the identified potential vulnerability. Tools like Metasploit are helpful in organizing an exploit's execution and functionality.
- Post-exploitation (reassessment) Phase
  - This phase focuses on assessing the value of the compromised machines, scanning the network further, and maintaining the accesses already gained. It can involve privilege escalation and finding internal network connections. It is a crucial step to understand how the adversary can locate the capabilities and vulnerabilities of the target system without detection and data is gathered from system information, moving laterally and scanning the network, searching through the file system(s), and using tools like Meterpreter modules.

- Reporting Phase
  - Involves summarizing all findings and steps taken and the recommended course of action to remediate the problems found.
  - Clear documentation and reporting assist in remediation.

In the sections below, we will cover each step's goals, the required information, and skills (thinking), and, if applicable, mention some tools used for discussion later on. It is important to remember that some steps can be performed independently and not part of an entire test; for example, network scanning and vulnerability assessments can be part of a regular workload for security teams.

Performing the steps and assessments regularly increases the data gathered. It improves the overall security posture by detecting threats more frequently, but it can be manual and labor-intensive. Automating the scans and the initial analysis, categorization, and prioritization of these scans should be a priority. One potential force multiplier is feeding this data into AI/RL tools, which could help maintain baselines and make it easier to automatically observe an increase in volume or a particular type of vulnerability.

2.1.5.4 Planning and Reconnaissance Phase (aka Information Gathering)

Planning and reconnaissance is the initial phase in penetration testing. The planning and reconnaissance phase aims to gather as much information about the target using as many sources as possible while remaining passive. Passive penetration testing refers to a state of investigation that does not involve any active scanning or risk of notifying the target that they are under investigation. The more active phases, which involve active interaction with target networks or personnel, take place after careful planning and reconnaissance. Proper planning and reconnaissance ensure all legal, ethical, and security considerations are in place. No one wants the authorities to investigate an intentional and legal activity, and no one wants to be the one to accidentally shut down a production system and cost a company or client money.

Once a team has a target or goal, they will gather information about the target systems, networks, organizational structure, and key personnel. This information can provide valuable insights into the target's organizational structure, what online systems are in use and their potential vulnerabilities. This information draws a complete picture of the target while sparking ideas for possible insecure entry points. The information gained during the reconnaissance phase is collated and used to develop a plan of attack that avoids detection.

The information gleaned during the recon phase comes from a variety of sources. Once this data is collected (and possibly during the data collection), machine learning can assist by providing a way to analyze the data quickly while making connections and predictions that might not be immediately apparent to a human. Common sources of information include looking through:

- **Publicly Available Information:** Websites, social media profiles and pages, public records searches, news archives, and good old search engines.
- **WHOIS Databases:** Internet registration records can provide information about a companies domain. This data can include contact details, names, addresses, registration dates, and name server information. Historical records can show temporal changes that allow a more complete history of an organization's digital footprint.
- **Domain Name Servers (DNS) Enumeration:** DNS keep records of IP to URL translations, thus revealing where infrastructure is hosted online. The records can also contain other useful information, such as subdomains and alternative IP addressing schemes where specific servers (mail, name, web, etc.) are hosted. This information highlights potential targets or points of entry.
- **Open Source Intelligence (OSINT)**: OSINT is the knowledge gleaned from collecting, collating, and analyzing information that comes from public sources. The information is usually used for specific purposes or to answer certain questions. OSINT can reveal more information then intended about a target, especially when combined with other intelligence sources used during reconnaissance. The information is mostly located in online repositories or databases, websites, directories, or unintentionally explanatory (oversharing) websites. It is usually obtained with little to no cost, but there are companies that collect information to sell to researchers as well.
  - OSINT tools are designed to assist the penetration tester in gathering and analyzing information from public sources.
  - An excellent primer on all things OSINT is Rae Baker's book, "Deep Dive" [35].

Another technique for performing reconnaissance is social engineering. It is not a passive step, but it can still be used during recon if the engagement goals and considerations allow it. Social engineering is a deceitful art form of targeting individuals or people in an organization using everyday communications [41]. The target is tricked into giving information valuable to the penetration testers that may not seem important to the company at the time. Using this information combined with other intelligence sources can reveal areas of weak security planning, unmask sensitive information or procedures, or even let attackers know the limits of security protocols. Potential avenues for social engineering attacks include phone calls, text messages, emails, or even physical, in-person interaction.

Red teams use network scanners to gather information about the network endpoints. Tools such as Nmap and Nessus provide information about open ports and running services. This information can highlight operating system versions, communications banners, and what protocols the port is operating on. This information is valuable when building network maps (such as needed for programming RL network topologies). It also can eliminate vulnerabilities for targeting by checking the research material for known vulnerabilities.

Scanning the network's IP space and ranges can expose the internal network attack surface. The engagement will occasionally require a passive scan to stay within the scope or bounds of the customer's goals. Active scanning can alarm SOCs and other intrusion detection measures that something is happening. A passive scan is non-invasive and generates no traffic on the customer's network. Thankfully, many online companies offer free and paid access to the data they have collected. They offer this information by continuously scanning the entire internet. Companies such as Shodan and Censys operate vast scanning machines that scan every IPv4 address on the internet multiple times a day and a good portion of IPv6 space. The data is also archived and can be searched historically for a temporal understanding of network configuration changes [24, 25]. The traffic generated by such companies is known as internet noise, and, as a result, many security platforms have been programmed to ignore them and their incessant chatter/requests [21, 30].

The penetration tester can use this data without fearing detection. No additional scans will be observed and seem unusual to the engagement's customer. Using the data they collect during their scans allows information collection by interacting directly with a target's systems.

Active scanning may occur later, but passive network data can reveal open ports and protocols, services operating on nonstandard ports, and SSH and SSL certificate information. It also provides an excellent way to passively map a target's network range and uncover which ranges are used for different purposes. For example, a company may own several IP blocks and separate out the addresses with the blocks for specific purposes. One block may be mail servers, one may contain websites, and the other may have a developer network.

2.1.5.5 Scanning

The scanning step is active, where penetration testers use tools, techniques, and potentially some machine learning assistance to find information about the target network or organization. Information of interest during this step includes any technical details related to the infrastructure that can be used in the following phases or to enrich existing data. At this point, the penetration team actively sends data to the IP ranges, netblocks, and URLs identified in previous steps. The scans can be targeted or general in nature. Targeted scans are performed to improve speed, and an initial assessment is completed to see if there are active listening boxes on each IP. The responding (alive) boxes are then probed on the well-known ports to test for services running with known vulnerabilities or configurations.

The targeted scanning is done initially since it will typically hit 100–400 ports instead of 65,535 possible ports per IP. Once there is an initial understanding of the target's footprint, a deeper scan occurs. This is also called a general or broad scan. This deeper scan expands the range of ports that are probed and expands the number and variety of protocols as well. For example, instead of just probing for quick-responding TCP-based network connections, a red team may take the time to probe for slow and potentially unresponsive UDP-based connections. These decisions and progressions of scanning actions will have been planned in the planning phases.

2.1.5.6 Port Scanners

One of the tools in the red team toolbox is the port scanner. Port scanners are programs or scripts designed to send specially crafted network packets to a target to discover which ports on an IP are listening. These ports are ready for action and responding to network connection requests. There are various tools, depending on the operator's preferred OS and experience level. The port scanner selected will also vary depending on the complexity of the scan desired, the level of data returned or needed, and the speed each tool takes to complete the scans. The most popular port scanners and where to find how to use them are:

- Nmap – nmap.org
- UnicornScan – kali.org/tools/unicornscan
- AngryIP scanner – angryip.org
- Netcat – netcat.sourceforge.net

Data collected from port scans drives the intelligence-gathering parts of the red team penetration tests, but it can also be collected for AI data gathering. Open ports are a vital part of making a network topology graph that is useful and valid (Figure 2.5) [36].

2.1.5.7 Network Discovery Tools

Network discovery tools are software that help penetration testers investigate the types of devices and traffic patterns on a given network. They are usually used from inside the network or, in the red team's case after an entry point has been found and exploited. Network discovery tools can automatically scan, map, and enumerate services and devices inside a network. While they have legitimate purposes, such as network troubleshooting and management, they are also accommodating for the red team. The information provided by these tools can identify servers, routers, switches, and endpoints while also providing details on OS, open ports, and services running in the network. Many of the port scanners listed above are used for this purpose. There are also paid software versions, like [15].

Port number	Service	Description
22	SSH	Secure shell for secure remote access
80	HTTP	Hypertext transfer protocol (web browsing)
443	HTTPs	HTTP over TLS/SSL (secure web browsing)
445	SMB	Server message block (Windows file sharing)
3389	RDP	Remote desktop protocol (Windows remote access)
53	DNS	Domain name system (name resolution)
139	NetBIOS	Network basic input/output system (legacy)
3306	MySQL	MySQL database server
1433	MSSQL	Microsoft SQL server
389	LDAP	Lightweight directory access protocol

Figure 2.5 Top used ports on the Internet.

2.1.5.7.1 Web Application Scanners Web application scanners are specialized software written to assess browser-based applications or interfaces' security levels (or lack thereof). For example, web pages that users log into or do everyday daily tasks on. In today's modern internet environment, almost every web page is a simple web application. However, a web application's complexity and level of service can go as high as using remote desktops or other software application suites (office email, word editing, spreadsheets, banking, etc.) right from the browser window.

Web application scanners are programmed to test the most well-known vulnerabilities and security holes for insecure cookie storage, directory traversal, SQL injection attacks, cross-site scripting vulnerabilities (XSS), and session management issues. Any of these found weaknesses could lead to compromise, aka a foothold, and a place for the attacker or penetration tester to pivot from. There are open source and paid versions of these valuable tools available, including:

- OWASP Zed Attack Proxy (ZAP) – https://www.zaproxy.org/
- Burp Suite – https://portswigger.net/burp
- Netspaker (now invicti) – https://www.invicti.com/

Information from web application vulnerability scanners is fundamental to the red team's arsenal. It is used for researching and finding vulnerabilities to test and exploit. It is also a practical data point in building RL systems. These systems may assist in analyzing large volumes of network data and quickly identify potential vulnerabilities, weak points, common threat entry points, weak protocol usages and patterns, or suspicious activity. It can use historical scan data to predict where future vulnerabilities might pop up.

2.1.5.8 Vulnerabilities Assessment Phase

After initial vulnerabilities are identified and thoroughly researched, the penetration testing team uses the data gathered from the reconnaissance and scanning phases to identify, categorize, and assess each facet in detail to assess its potential severity and impact on the target system's security posture. The vulnerabilities are analyzed individually and in groups to assist in coming up with a plan of execution for the exploitation phase.

While most of the assessment relies on the penetration tester's skills and experience, several industry-standard frameworks, and/or tools are often used to speed up the process and ensure it is completed thoroughly. A combination of automated tools and manual methodologies scrutinizes the network's security posture, attached systems, and running services. Pentesters are looking for potential loopholes and theoretical vulnerabilities that manifest into unintentional or unauthorized access during this phase.

The usual series of steps taken to assess a vulnerability is scanning for the vulnerability and detecting a vulnerability. Once the vulnerabilities are all detected, or at least as many as possible, they are prioritized for their applicability and exploitability potential. There may be hundreds or thousands of identified vulnerabilities, so assistance from AI technologies is a blessing. The prioritized list of vulnerabilities to try is then checked for the possibility of false positives. This process repeats until the red team has a comfortable number of vulnerabilities to actively test and exploit in the next phase. As mentioned, time is short, and data can get large.

Thoughts on AI Assistance: AI/machine learning technologies can assist in this step by automating the assessment process, prioritizing the vulnerabilities based on their potential risk and impact, and potentially suggesting remediation methods.

- Vulnerability detection – AI algorithm can be trained to detect vulnerabilities in the target system by analyzing historical data and identifying patterns and potentially vulnerable points.
- Prioritization of vulnerabilities? Using known scores and historical data compared to data revealed during the scanning phase, can the most likely successful vulnerabilities be put on top and potentially lesser ones assessed for new methods?
- Reduce false positives from scanning data by combining multiple sources and reducing the rate of false positives.

- Assess prioritization and value? Creating lists of assets that are more critical to company infrastructure, allowing focus on the most important first (saving time of tester).

2.1.5.9 Exploitation Phase

Data has been gathered, collated, analyzed, assessed, and prioritized. The penetration testing team is ready to check the plan and start taking action! Exploitation is the pentesting step where the tester attempts to use the identified vulnerabilities and gain "unauthorized" access to or control the target systems or networks. Probed ports that returned vulnerable-looking banners are more thoroughly examined and tested. Protocols researched and found to potentially contain vulnerabilities are documented and leaned on. Mail servers, DNS servers, web servers, routers, and firewalls are all possible access points. Another likely scenario is a situation revolving around the successful execution of a fishing campaign.

Once an exploit has been delivered and detonated by the "target," the red team continues to the next phase as the jumping-off point for repeating the cycle. Scanning the network and gathering information, researching the potentially exploitable services, and looking for a place to maintain access, escalate access, pivot in the network, or move laterally throughout and explore to find more vulnerabilities (Figure 2.6).

The team must use the tools they have to follow the plan of action responsibly. As discussed earlier, the object of the penetration test is not to actually cause any harm to the target systems, just to assess their weaknesses and vulnerabilities and report the findings. This includes damages caused by unintentional actions.

Thoughts on AI Assistance: AI assistance would benefit a red team during this phase by potentially analyzing historical data and identifying attack patterns that indicate potentially lucrative attack and exploitation techniques. It could also aid in attack path mapping and highlighting weak network access points. It is difficult

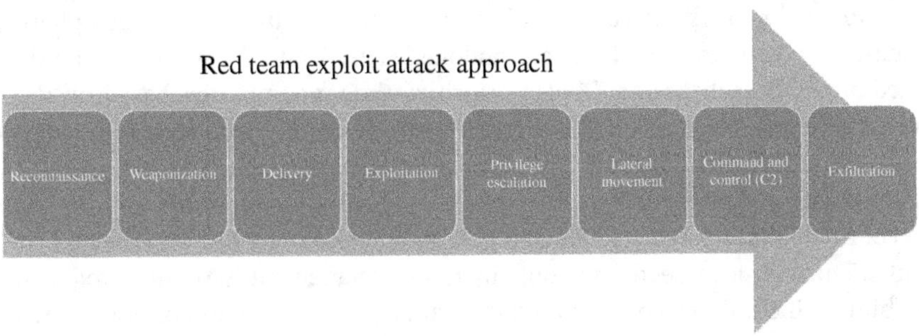

Figure 2.6 Exploitation paths.

to know where to pivot in a network and having an AI system that assisted in predicting the success rate of potential action actions would make strategies more efficient. This is especially true and important if a "quieter" test is desired and needs to leave as little trace and/or make as little noise as possible. System logs and other evidence of user behaviors (such as network traffic) can be collected during this and previous phases. These logs can be pretty significant and having technology to assist in analyzing the records for patterns would reduce the time for analysis and action.

2.1.5.10 Post-exploitation (Reassessment) Phase

The post-exploitation phase is the phase where the penetration team assesses the successfully executed exploits for completeness, performs further scanning and vulnerability research, explores how to maintain the exploit and presence, and tries to expand their presence through other exploits and network exploitation (also known as pivoting). The primary purpose of this phase is to maintain a persistent presence, avoid detection by the blue team (if present), and determine the capabilities and value of the target system and networks. The testers continue scanning and exploiting methods, collecting data on the target organization, and its networks until the team achieves its goals.

2.1.5.11 Reporting

This final phase is where the penetration testing team summarizes all the findings in a detailed report that includes all the vulnerabilities and recommendations for remediation and prevention. The fun doesn't stop once all the testing is done! The findings must be organized, summarized, and presented to the security teams and leadership understandably so they can comprehend the results and start forming plans to take action, often based on the red team's findings and recommendations.

The customer's security teams can use the report generated during this phase as a guideline for remediating any network and system vulnerabilities to improve the company's security posture and prevent future attacks.

The report includes the vulnerabilities discovered during the previous phases, the data found and exploited, the methodologies used, and the success of the simulated attack/breach. It should also prioritize the vulnerabilities based on their severity, potential impact on the target systems, and the likelihood of attackers using them for an actual attack. (AI can potentially assist. See below.)

Along with a high-level overview report of the findings suitable for a C-Suite level or nontechnical staff, a detailed report should also be included in the reported results. The high-level sections should include a detailed outline of uncovered vulnerabilities, including common/industry-standard metrics (like common vulnerability scoring system [CVSS] scores), a business impact assessment to demonstrate the potential risks, an executive summary, and details on the recommendations for remediation. It is often said that the reporting phase is the most critical phase for everyone involved since it not only contains the information found during the penetration test but is highly likely to be read by the technical IT staff and senior management.

An outline of information typically included in reports for cyber operations is detailed below:

* High-level overview * Summary of Vulnerabilities found - Detail outline - CVSS scores * Impact assessment - The how and where it hits the network, and also the business * Technical details * Information/data found - Sources and methods - E.g. OSINT on facebook or Linkedin * Remediation recommendations - Risk mitigation strategies - Technical details - Contextualization - Summary * Business risk statement – this ties the technical data to factual business data and makes the technical data more comprehensible by the C-Suite.

Aligned with report generation is the utility of AI for cyber analyst work flows. AI can support automatically generating report data, trends, or issues based on crucial findings and provide visualizations for summaries. In addition, models can be trained to prioritize remediation actions based on historical impacts and the most effective solutions. Using these types of technologies are critical to helping to assess the likelihood, potential impacts, and success rates of discovered vulnerabilities (based on historical data? Large data?)

2.2 Importance of Data

Data is an essential part of red team penetration testing. Data is even more critical when designing and implementing machine learning solutions to assist the red team's efforts. Collecting and analyzing data points is second nature to red teams and is worked into every step in the process/methodology. Data can be found almost everywhere you look, and this section will detail some of the data sources red teamers use to perform their duties.

Machine learning and RL require prerequisite knowledge of specific data points, behavior expectations, and available options. More importantly, performing the desired analytics properly requires lots of data. Data is collected by running different scanners and security tools and analyzing/comparing using databases and online search resources. It is essential to understand that data comes in many forms and must be extracted, transformed, and loaded (ETL'd) into formats/tools/repositories that the machine learning models can understand and use during their processing and calculations. Working closely with red team practitioners during the model design process and understanding how the data is used during a penetration test results in better-performing models and accurate outcomes.

2.2.1 Data Types, Data Sources, and Pivot Points

Understanding data is essential due to the dynamic nature of penetration and the diverse nature of the data collected. Data can come from many sources and be used as jumping-off points (pivot points). Taking data found from one source, analyzing it for relevance and applicability, and then searching the data in other sources is the lifeblood of the red team analyst. This section will detail the sources of relevant data. It will also list potential tools and provide links if they haven't been discussed. It will also contain examples of how data may be structured (or unstructured) and require custom scripts for practical use.

2.2.1.1 Scanning Data via Port Scanners and Network Scanners

A port scan is a technique used to identify which ports are open on a computer or network device. It involves sending client requests over the network to a range of server port addresses to determine if any services are listening (waiting for client connections) on that port. Red teams analyze the response data (or lack thereof) to make several assessments and inferences. Port scanning is also used for legitimate purposes, such as network diagnostics, security assessments, and endpoint information gathering (endpoint mapping). Due to the usefulness of network scanning, it is rarely fully disabled on a network, which makes it a valuable tool for the red team information-gathering arsenal. Several open-source and commercially available tools can be used to gather port-related information. Some of the most popular in use today are:

- Nmap – nmap.org. Nmap is among the most popular, oldest, and free tools for port scanning and enumeration. It has many extensible and customizable libraries (also known as modules) called NSE libraries [16] that can extend functionality beyond simple port scanning [16].
- Massscan [11]. Massscan is a fast and noisy scanner that can, according to claims on its GitHub page, scan the entire internet in under 5 minutes, transmitting 10 million packets a second from a single machine. This tool is incredibly noisy and would not be used in networks where stealth is desired. Instead, it quickly helps get a broad sense of an external network range.
- Zmap [33]. Zmap is another fast but noisy internet scanner that can send single packets to network endpoints to get a network footprint quickly. As with Massscan, the tool is not designed for stealth but for scale and speed.
- Unicornscan [19]. Unicornscan is a free, open-source scanning tool commonly installed on Kali Linux. Its advanced capabilities and custom module support make it popular with red teams/penetration testers, making active connection tests, automatically identifying OS, and logging the data in PCAP or database formats.

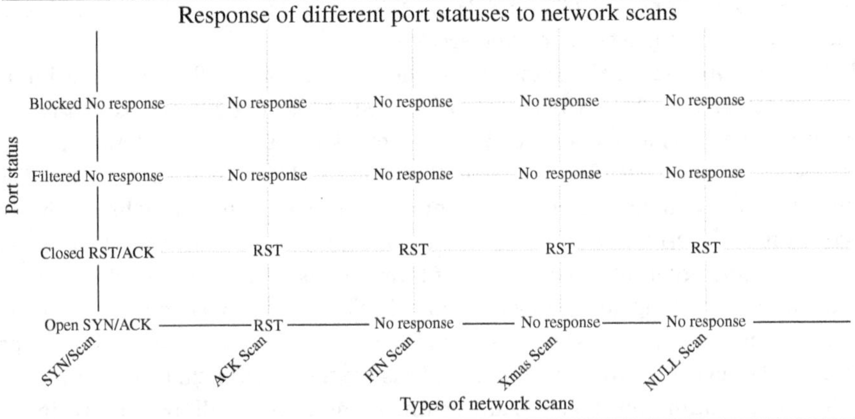

Figure 2.7 Ports responding to scans, revealing their state.

Data received/collected during port scanning can indicate opportunities to reveal open ports by capturing the data on ports that respond positively to the scanning utilized. This means that the port on the server actively responded to the scan and is waiting for connections (legitimate or not) to be made to provide the requested services. This information is collected by the red team to determine which network services are running on the specific servers/devices connected to the network. In addition, the same port scan can also reveal closed ports, also known as denied. This information indicates that no servers are listening on those ports, and no data flows to or from the servers/devices. This information may appear unuseful at first, but it can be used to understand the network configurations and router settings and identify potential security measures that are in place. Discovering ports that are filtered for one reason or another is also possible. Filtered ports differ from closed ports in that where a closed port can be confirmed as nonactive, a filtered port may be unreachable due to filtering or security measures such as a firewall, gateway, or other active security measure. Noting the filtered port data can reveal the presence of these security measures and allow the red team to take appropriate steps during the planning or execution phases (Figure 2.7).

2.2.1.2 Application Identifications (Banners)

Port scanning can reveal the open ports and potentially what services or applications are present on a network endpoint's system. Typically, applications used in corporate networks adhere to standard configurations to avoid conflicts and allow easy configuration and setup of network services. In fact, the range of ports from 0 to 1024 are known as "registered ports" and are reserved for known network services. For example, port 22 for SSH, port 80 for HTTP, port 443 for HTTPS, etc. (See Appendix for a list of ports and services.)

Services running on nonstandard or uncommon ports will also return information to the requester when the port is open. These open ports excite the red team because it is a customized server. Humans make errors and creating custom settings is a great place to locate these errors. The information returned is known as the banner. Banners can contain information that is highly useful to red teamers. Information includes service name, version number, product name, port used, and other information. This information is considered valuable enough that banner-grabbing attack tools and tutorials exist. The most common banner-grabbing attack techniques are passive and active. Some of the more exciting things that can be revealed through banner analysis are the services that can be found.

Services running on nonstandard or uncommon ports will also return information to the requester when the port is open. These open ports excite the red team because it is a customized server. Humans make errors, and creating custom settings is a great place to locate these errors. The information returned is known as the banner. Banners can contain information that is highly useful to red teamers. Information includes service name, version number, product name, port used, and other information. This information is considered valuable enough that banner-grabbing attack tools and tutorials exist. The most common banner-grabbing attack techniques are passive and active as in the preceding, apparently duplicate paragraph. Some of the more exciting things that can be revealed through banner analysis are the services that can be found.

2.2.1.3 Operating System Identification

Identification of the operating system running on the network device can also be assessed using port scanning. Each operating system typically has a set of default protocols installed for responding to network requests. A red team can collect the responses received during a scan, compare this information to the known response formats, and assess which operating system fits the profile. This will allow them to scope and plan the phases as the data is received. For example, windows machines will respond with a (close) signal to ports that are not open on the system, whereas certain Linux machines will just refuse to answer and drop the network connection.

2.2.1.4 Network Topology

Lastly, port scanning can provide valuable insights into the network topology. Not only can it reveal which IP addresses have active machines on them, but it can also reveal the presence of security devices like firewalls or proxy servers. The responses received during a port scan help identify the network infrastructure pain points and potential points of entry for the red team attackers.

An essential part of the collection process, saving the data returned, plays a prominent role in applying the trade to AI technology and assistance. Collecting and formatting this data into machine-readable network maps allows machine learning technologies to feed the network topology into the model. The model can then combine the information with all the other data points, such as known vulnerabilities, CVSS scores, CVE indicators, etc., to drive the model to better attack path generations, exploit execution priority(ies), and other valuable information that will shorten the time to target and improve the chances of successful attack execution.

2.2.1.5 Network Scanners

Automated network scanners are used by penetration testers to perform vulnerability scans automatically and identify potential security weaknesses on a network or system. Using these computerized tools allows the process to be sped up and automated, but the setup and analysis of the returned data still rely on manual inspection and understanding. Network scanners can also identify many vulnerabilities across various assets, including servers, laptops or desktops, firewalls, printers, containerized systems, etc. Using these scanners can help regularly monitor networks using scripts, can assist in reporting and remediation documentation, and can be part of a compliance requirement set by industry or government regulations to maintain and improve an organization's security posture. However, most of this process and data analysis still rely on the human analyst's interpretation and decoding, so machine learning has a huge opportunity to increase this process's efficiency, speed, and overall cost. Experienced network analyst professionals are expensive.

Available in commercial and open-source variants, several automated vulnerability scanners are used to identify network vulnerabilities. Several popular ones are listed below:

- **Nessus:** Nessus [15] is a widely used vulnerability scanner that can perform comprehensive network scans to identify system, application, and network device vulnerabilities. It provides detailed reports and prioritizes vulnerabilities based on severity. See the data section below regarding output formats and considerations.
- **Qualys:** Qualys [29] offers a suite of vulnerability management solutions, including a network vulnerability scanner. It scans networks for vulnerabilities, misconfigurations, and compliance violations. It provides real-time visibility into the security posture of the network.
- **Rapid7 Nexpose:** Nexpose [27] is a vulnerability management solution with a network vulnerability scanner. It scans networks, identifies vulnerabilities,

and provides actionable insights for remediation. It also integrates with other security tools for a comprehensive security program.

- **Open Vulnerability Assessment System (OpenVAS)**: OpenVAS [28] is an open-source scanner that scans networks for known vulnerabilities. It provides a web-based interface and generates detailed reports.
- **Tenable.sc:** Tenable.sc (formerly SecurityCenter) [32] is a vulnerability management platform with a network vulnerability scanner. It scans networks, identifies vulnerabilities, and provides risk scores to prioritize remediation efforts. It also integrates with other security tools for a holistic security approach.

2.2.1.6 External "Passive" Data Sources

Plenty of data and intelligence can be gathered without making a connection or sending a packet to the target. An attacker often prefers not to generate any active traffic, especially in real-world scenarios. Noise-generating, heavy traffic, such as actively scanning a target's network during the reconnaissance phase, can alert the target to the activities and inform them that an attack may be on the way.

Passive data sources are rich sources of information for both an attacker and a penetration tester. They are helpful for data points to feed into the RL models for training. The data is considered passive when it is already collected by third parties and stored in databases for retrieval. These third-party collection platforms are constantly scanning the internet for information. They are, therefore, usually lost in the noise of the internet or actively ignored by security teams due to the large number of alerts they would receive otherwise. A red team can take advantage of this gathered information while remaining quiet during the reconnaissance phase and get a good picture of what the external network connections look like at a point in time.

Some of the most common sources of passive network data include services like Shodan.io [30], Censys.io [21], and greynoise.io [23], among others. Not only is network information available from external sources, but valuable registration information may be available from sources like DNSDumpster.com

2.2.1.7 Databases

At the time of this writing, over 220k [14] total vulnerabilities are listed in the national vulnerability database. With the number of vulnerabilities continuously growing and the number of connected devices on the Internet, the problem isn't getting any smaller. Add in the fact that newer technologies are continually being developed, novel uses for data flowing on the Internet are being found, and the proliferation of automated machines and cloud-based robotic research (GPT LLM type stuff) exchanging information on the Internet is also expanding. There are a lot of existing and potential points of vulnerability to be considered. With all this threat and vulnerability data, it seems almost impossible to figure out or even understand where to start sorting and classifying it. Machine learning is a solution that can be leveraged and explored.

2.2.1.8 Vulnerabilities Databases: CVE and CVSS Databases

Vulnerability databases are extensive collections or repositories of data about identified security vulnerabilities. They can be collected and sorted based on hardware or software. These databases are excellent resources for red teams because they contain detailed insights and technical information surrounding vulnerabilities. The data includes descriptions, severity scores, operations impacts, and potential solutions or workarounds. Some of the critical databases discussed below are used by red teamers and network security professionals worldwide. The bad guys use them too.

CVE and the CVSS are two widely used sources of information to document and define known vulnerabilities. CVSS is a free and open industry standard for assessing the severity of system security vulnerabilities [20, 29]. It uses a scoring system based on 0–10, with 10 being the most severe, and allows system defenders and network analysts to prioritize responses, resources, and attacks in the case of the red team. The CVE program, on the other hand, is a reference method for publicly known and disclosed information-security vulnerabilities and exposers. It provides a dictionary of identifiers that can assist both blue and red teams in assessing and referencing necessary patches or red team findings. It was launched by the MITRE corporation in 1999 [26] and is still widely used. CVSS was developed by the National Infrastructure Council in collaboration with the NIST [20]. Both are essential tools for red teams as they provide a standardized way to identify, assess, and prioritize vulnerabilities and findings during testing.

CVSS scores can be used along with CVE findings in the RL process as a pivot point or data point for decisions. During attack graph generation, it can be used to represent the potential attack paths and vulnerabilities that exist in a network. CVE parameters, which include specific information about specific vulnerabilities, can be used as input nodes for the attack graph.

CVSS scores can be used along with CVE findings in the RL process as a pivot point or data point for decisions.

During attack graph generation, it can be used to represent the potential attack paths and vulnerabilities in a network. CVE parameters, which include specific information about specific vulnerabilities, can be used as input nodes for the attack graph.

The combination of CVE and CVSS has been used as parameter and characteristic definitions for attack graph analysis in IoT and IT infrastructure. Mining the data for information and severity allows the attack model to represent severity and potentially unmitigated risks, allowing the natural progression to using RL or other AI models to enhance the reliability and speed of attack graph generation while reducing the overall human-analytical costs [34].

2.2.1.8.1 Github [22] GitHub can be a goldmine for penetration testers. Many security researchers and developers publish their security-related tools, scripts, and resources on GitHub. Searching for terms like "pentest," "security," or specific technologies can yield valuable tools and resources.

GitHub can contain helpful tools and scripts and be used as a data source during a pentest's reconnaissance phase. Frequently, companies use GitHub as a code repository for their development. Developers may leave clues or data lying around without realizing it.

2.2.1.8.2 Emails/Compromised Credentials Email addresses provide a valuable point of information for red teams. Care should be taken to monitor emails going into and out of the network to ensure none of the methodologies listed below are being used to target users, executives, or administrators.

Phishing – Red teams can use emails to send targeted phishing emails to potential targets that operate inside the network. These emails can be links to malicious content or crafted to trick the target into revealing sensitive information.

Finding compromised passwords – Emails are often used to register a user to a website, and if that website has had a breach or leak, that information may be found and used by the penetration tester. Websites that collect and allow searching of breach data, like haveibeenpwned.com [24], enable searching, and organizing email data easier.

Social engineering – Email addresses can be used in social engineering attacks where the red team impersonates legitimate individuals or organizations to manipulate targets into revealing confidential information or performing actions that allow the attacker to benefit.

OSINT gathering – An email can be the starting point, or seed, for an OSINT gathering operation. Once emails are found, they serve as pivot points into different databases for further exploitation.

Credential stuffing – Email addresses in breach data dumps can be used in credential-stuffing attacks. These attacks rely on the tendency for users to reuse passwords across multiple accounts and services. This password reuse is even more likely if legitimate emails within the target's domain are used for registration.

2.2.1.8.3 Hunter.io – Finding Emails Hunter.io [25] is a web-based service that provides email finding and verification tools for businesses and individuals. The primary feature of Hunter.io is its ability to help users find email addresses associated with a particular domain or company. SalesIt's and marketing professionals, recruiters, and researchers often use it to identify and contact potential leads, customers, or collaborators. However, it is also used by red teams/penetration testers to get the company's email format (first_last@target.com, f.last@target.com, etc.), as well as find the company's C-Suite and technical personnel's email addresses. Consider how handy it would be to know who the CISO or chief IT security person is for targeting purposes.

2.2.1.9 Data Formats

Data comes in many formats from different scanners sources. Understanding how the data is returned from any given source is essential. The data must be cleaned and readable. This applies to humans and whether the data will be used in scripts, programs, or AI applications. Humans and computers alike can only use what they can read and understand. This section covers formats seen from various tools but is not all-inclusive.

For example, Nessus alone has eight output formats, as shown in the figure below (Figure 2.8). Understanding where the data will end up and how it will be used by the AI models before conducting the scans can help. However, it is only sometimes possible to get the data in the format you want from the security teams or scanners, so it is a good idea to understand the various formats and what steps will be needed to ETL the data for later use.

Table from (https://docs.tenable.com/nessus/Content/ScanReportFormats.htm).

To process the data into a format readable by the machine learning models, it is first necessary to obtain the data from the scans, logs, or databases as appropriate. The data must be converted into the desired format using scripts or automated programs. Once the data is extracted and transformed, it can be loaded into whichever RL systems are being developed.

Some examples of various data formatting are included below for consideration.

2.2.1.10 Nessus Typical File/Terminal Output

«< begin extracted code »> Vulnerability Title: Microsoft Windows SMB Remote Code Execution

(CVE-2021-34567) Severity: Critical Plugin ID: 12345 Plugin Family: Windows CVE ID: CVE-2021-34567 CVSS Base Score: 9.8 (Figure 2.9)

Description: This vulnerability allows remote attackers to execute arbitrary code on a target Windows system by sending a specially crafted packet to the SMB service. Successful exploitation could lead to complete compromise of the system.

Nessus	A .nessus file in XML format that contains the list of targets, policies defined by the user, and scan results. Nessus strips the password credentials so they are not exported as plain text in the XML. If you import a .nessus file as a policy, you must re-apply your passwords to any credentials.
Nessus DB	A proprietary encrypted database format that contains all the information in a scan, including the audit trails and results. When you export in this format, you must enter a password to encrypt the results of the scan.
Policy	An informational JSON file that contains the scan policy details.
Timing data	An informational comma-separated values (CSV) file that contains the scan hostname, IP, FQDN, scan start and end times, and the scan duration in seconds.
Reports	
PDF	A report generated in PDF format. Depending on the size of the report, PDF generation may take several minutes. You need either Oracle Java or OpenJDK for PDF reports.
HTML	A report generated using standard HTML output. This report opens in a new tab in your browser.
CSV	A CSV export that you can use to import into many external programs such as databases, spreadsheets, and more.

Figure 2.8 Screen of all Nessus reporting options (file/data formats). Source: From [31], 2023, Tenable, Inc.

Solution: Apply the latest security patch provided by Microsoft to address this issue. Ensure that the affected system's software is up to date.

Affected Systems: - Microsoft Windows 7 - Microsoft Windows 10 - Microsoft Windows Server 2016

References: - https://cve.mitre.org/cgi-bin/cvename.cgi?name=CVE-2021-34567 - https://blog.netwrix.com/smbv3-vulnerability

Proof of Concept: A proof-of-concept exploit has been publicly demonstrated. It is recommended to address this vulnerability as soon as possible.

«< end extracted code »>

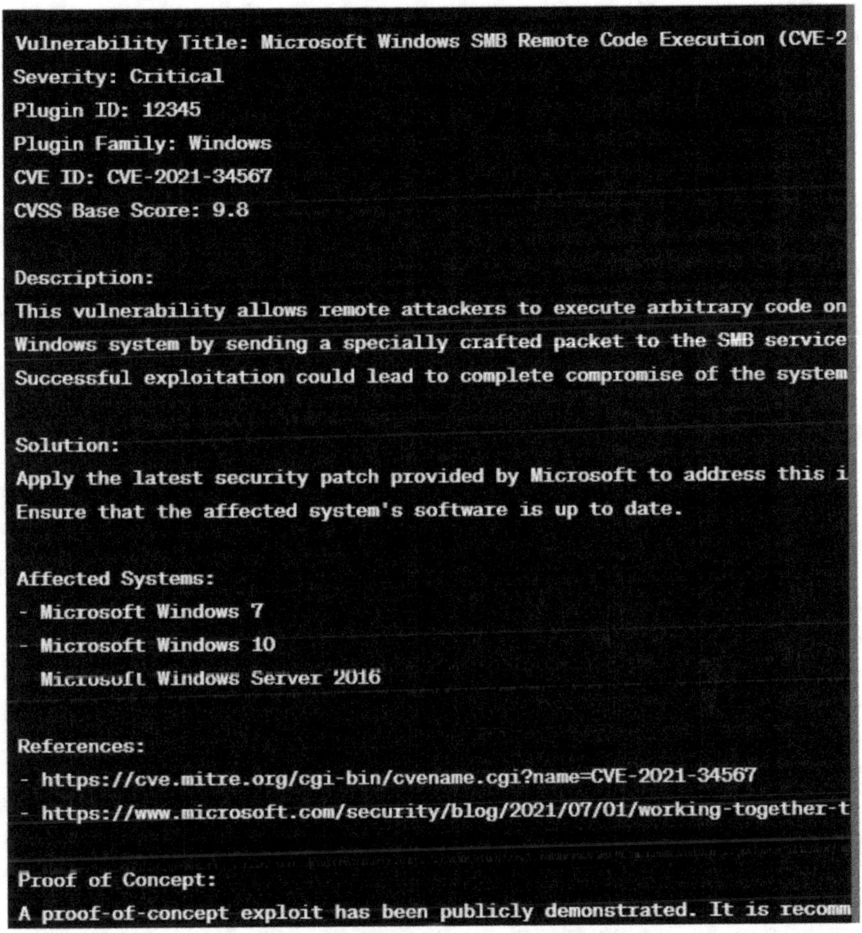

```
Vulnerability Title: Microsoft Windows SMB Remote Code Execution (CVE-2
Severity: Critical
Plugin ID: 12345
Plugin Family: Windows
CVE ID: CVE-2021-34567
CVSS Base Score: 9.8

Description:
This vulnerability allows remote attackers to execute arbitrary code on
Windows system by sending a specially crafted packet to the SMB service
Successful exploitation could lead to complete compromise of the system

Solution:
Apply the latest security patch provided by Microsoft to address this i
Ensure that the affected system's software is up to date.

Affected Systems:
- Microsoft Windows 7
- Microsoft Windows 10
  Microsoft Windows Server 2016

References:
- https://cve.mitre.org/cgi-bin/cvename.cgi?name=CVE-2021-34567
- https://www.microsoft.com/security/blog/2021/07/01/working-together-t

Proof of Concept:
A proof-of-concept exploit has been publicly demonstrated. It is recomm
```

Figure 2.9 Nessus normal text report format example screenshot.

This normal format is great for reading on a screen or command line interface (CLI) when the number of vulnerabilities is low. The format is not great for reading through and analyzing in bulk however. It is also suboptimal for using in any form of data processing or model development. The data scientists will have to write custom scripts to extract, transform, and load the data into useful services.

2.2.1.11 Nessus Output, CSV Format

«< begin extracted code »> Plugin ID, Plugin Name, Severity, CVE, CVSS Score, Affected Host, Description, Solution 12345, Microsoft Windows SMB Remote Code Execution, Critical, CVE-2021-34567,9.8,192.168.1.100, “This vulnerability

```
Result ID,Host,Port,NVT Name,Severity,CVE,CVSS Score,Description,Soluti
123456,192.168.1.100,80,Microsoft Windows SMB Remote Code Execution (CV
```

Figure 2.10 Nessus CSV report format example screenshot.

allows remote attackers to execute arbitrary code on a target Windows system by sending a specially crafted packet to the SMB service (Figure 2.10). Successful exploitation could lead to complete compromise of the system." "Apply the latest security patch provided by Microsoft to address this issue. Ensure that the affected system's software is up to date."
«< end extracted code »>

As seen in the screenshot and extracted text above, the CSV format may be a little less human readable, but more usable by data scientists for data ingestion. Importing CSV formatted files is a common approach to handling large amounts of structured data. This format is also easier to concatenate and combine for massive AI applications as well.

2.2.1.12 Nessus Output, XML Format

«< begin extracted code »> <Report> <Host> <HostProperties> <host-fqdn> example.com</host-fqdn> <operating-system>Microsoft Windows 10 </operating-system> </HostProperties> <ReportItems> <ReportItem pluginID="12345" pluginName="Microsoft Windows SMB Remote Code Execution" severity="Critical" cve="CVE-2021-34567" cvssScore="9.8"> <description>This vulnerability allows remote attackers to execute arbitrary code on a target Windows system by sending a specially crafted packet to the SMB service. Successful exploitation could lead to complete compromise of the system.</description> <solution>Apply the latest security patch provided by Microsoft to address this issue. Ensure that the affected system's software is up to date.</solution> </ReportItem> </ReportItems> </Host> </Report>
«< end extracted code »>

XML is a tagged format with parsers available in Python and other language, so it may be a useful format to use when there are existing ETL pipeline requirements. On its own it can be parsed and ETLed into the final destination easily in bulk (Figure 2.11).

2.2.1.13 OpenVAS Standard Format

«< begin extracted code »> Result ID: 123456 Host: 192.168.1.100 Port: 80 NVT Name: Microsoft Windows SMB Remote Code Execution (CVE-2021-34567) Severity: High CVE: CVE-2021-34567 CVSS Base Score: 9.8 Description: This vulnerability allows remote attackers to execute arbitrary code on a target Windows

```
<Report>
  <Host>
    <HostProperties>
      <host-fqdn>example.com</host-fqdn>
      <operating-system>Microsoft Windows 10</operating-system>
    </HostProperties>
    <ReportItems>
      <ReportItem pluginID="12345" pluginName="Microsoft Windows SMB Re
        <description>This vulnerability allows remote attackers to exec
        <solution>Apply the latest security patch provided by Microsoft
      </ReportItem>
    </ReportItems>
  </Host>
</Report>
```

Figure 2.11 Nessus XML report format example screenshot.

```
Result ID: 123456
Host: 192.160.1.100
Port: 80
NVT Name: Microsoft Windows SMB Remote Code Execution (CVE-2021-34567)
Severity: High
CVE: CVE-2021-34567
CVSS Base Score: 9.8
Description: This vulnerability allows remote attackers to execute arbi
Solution: Apply the latest security patch provided by Microsoft to addr

References:
- https://cve.mitre.org/cgi-bin/cvename.cgi?name=CVE-2021-34567
- https://www.example.com/security/advisory/CVE-2021-34567

Proof of Concept: A proof-of-concept exploit has been publicly demonstr
```

Figure 2.12 OpenVAS normal text report format example screenshot.

system by sending a specially crafted packet to the SMB service. Successful exploitation could lead to complete compromise of the system (Figure 2.12). Solution: Apply the latest security patch provided by Microsoft to address this issue. Ensure that the affected system's software is up to date.

```
Plugin ID, Plugin Name, Severity, CVE, CVSS Score, Affected Host, Descr
12345, Microsoft Windows SMB Remote Code Execution, Critical, CVE-2021-
"This vulnerability allows remote attackers to execute arbitrary code o
Successful exploitation could lead to complete compromise of the system
"Apply the latest security patch provided by Microsoft to address this
Ensure that the affected system's software is up to date."
```

Figure 2.13 OpenVAS CSV report format screenshot.

References: - https://cve.mitre.org/cgi-bin/cvename.cgi?name=CVE-2021-34567 - https://nvd.nist.gov/vuln/detail/CVE-2021-34567

Proof of Concept: A proof-of-concept exploit has been publicly demonstrated. It is recommended to address this vulnerability as soon as possible.

«< end extracted code »>

2.2.1.14 OpenVAS.csv Format

«< begin extracted code »>

Result ID, Host, Port, NVT Name, Severity, CVE, CVSS Score, Description, Solution, References 123456,192.168.1.100,80, Microsoft Windows SMB Remote Code Execution (CVE-2021-34567), High, CVE-2021-34567, 9.8, "This vulnerability allows remote attackers to execute arbitrary code on a target Windows system by sending a specially crafted packet to the SMB service. Successful exploitation could lead to complete compromise of the system.","Apply the latest security patch provided by Microsoft to address this issue. Ensure that the affected system's software is up to date."," https://cve.mitre.org/cgi-bin/cvename.cgi?name=CVE-2021-34567, https://nvd.nist.gov/vuln/detail/CVE-2021-34567"

«< end extracted code »>

These examples show how complex that data can be when it is returned (Figure 2.13). The data collected and analyzed during a red team engagement may come from many different sources. Data management, planning, and synchronization come into play as the engagement continues and the volume of data increases. It is essential to realize all this data must be ETL'ed eventually, so don't let it get out of hand. Data collection solutions may exist, but future development into creating an AI system that can ingest, transform, and analyze the data will benefit the community.

2.3 Conclusion

This chapter focused on introducing the reader to red team and penetration concepts. It started by explaining the importance of penetration testing for many industries. It continued with defining the various types of teams that can be found in the security industry and what they are responsible for. A brief history of red team operations and where the concepts stemmed from allowed the reader to understand that the ideas predated computers and had been evolving for many decades. We then reviewed the main objectives and considerations that red teams consider before, during, and after an engagement.

The chapter continued by listing, then detailing, the methodology and phases that occur during a penetration test. Each phase was explored to give the reader an idea of how the red team operates and thinks, what data may be available, what data may be collected, and where tools and research must be used to pivot and make critical decisions. All these steps are performed to find, report, and remediate vulnerabilities.

Once the reader understands the data collected during the penetration testing phases, the chapter gives examples of where additional data can be collected. This data is helpful for the penetration test and development of AI solutions and tools. Because data comes from so many sources and is rarely standardized across tools and industries, the chapter provided examples of different data formats so the reader can generate ideas on how to process the data at a future time. Emphasis was placed on the importance of having a data strategy and the criticality of the ETL pipeline.

References

1 U.S. Cyber Command History. URL https://www.cybercom.mil/About/History/.

2 A product of the TRADOC G-2 operational environment enterprise: the red team handbook. URL https://nsiteam.com/social/wp-content/uploads/2017/01/RTHB_v7.0_Web.pdf.

3 Pentest execution standard, Aug 2014. URL http://www.pentest-standard.org/index.php/Main_Page.

4 PCI DSS Standards, Jul 2018. URL https://listings.pcisecuritystandards.org/documents/PCI_DSS-QRG-v3_2_1.pdf.

5 Open source security testing methodology manual: contemporary security testing and analysis, 2019. URL https://www.isecom.org/OSSTMM.3.pdf.

6 Electronic health record systcms, Feb 2020. URL https://www.hhs.gov/sites/default/files/electronic-health-record-systems.pdf.

7 Penetration testing methodology, Dec 2023. URL https://www.getastra.com/blog/security-audit/penetration-testing-methodology/.

8 Cyber security guidance material, Jul 2023. URL https://www.hhs.gov/hipaa/for-professionals/security/guidance/cybersecurity/index.html.

9 NiST. Oct 2023. URL https://www.nist.gov/cybersecurity.

10 U.S. Cyber Command. Joint cyber operations strengthen DoD networks, Oct 2023. URL https://www.cybercom.mil/Media/News/Article/3565190/joint-cyber-operations-strengthen-dod-networks/.

11 Masscan, 2023. URL https://github.com/robertdavidgraham/masscan.

12 Dictionary definition of methodology, 2023. URL https://dictionary.cambridge.org/us/dictionary/english/methodology.

13 National Institute of Standards and Technology (NIST), Dec 2023. URL https://www.nist.gov/.

14 National vulnerability database, Dec 2023. URL https://nvd.nist.gov/general/nvd-dashboard.

15 Nessus, 2023. URL https://www.tenable.com/products/nessus.

16 NMAP.ORG, 2023. URL https://nmap.org/nsedoc/lib/nmap.html.

17 Open worldwide application security project (OWASP), May 2023. URL https://owasp.org/www-project-top-ten/.

18 NIST. Payment card industry data security standard, 2023. URL https://csrc.nist.gov/glossary/term/payment_card_industry_data_security_standard.

19 Unicornscan, 2023. URL https://www.kali.org/tools/unicornscan/.

20 What is the common vulnerability scoring system (CVSS), May 2023. URL https://www.sans.org/blog/what-is-cvss/.

21 Censys search engine, 2023. URL https://www.censys.io/.

22 Github.com, 2023. URL https://github.com.

23 Greynoise.io, 2023. URL https://www.greynoise.io/.

24 Haveibeenpwned.com, 2023. URL https://haveibeenpwned.com/.

25 hunter.io, 2023. URL https://hunter.io/.

26 Common vulnerability and exploits (CVE) resources and support, 2023. URL https://www.cve.org/ResourcesSupport/FAQs.

27 Nexpose, 2023. URL https://docs.rapid7.com/nexpose/.

28 Openvas, 2023. URL https://www.openvas.org/.

29 Qualys, 2023. URL https://www.qualys.com/.

30 Shodan.io, 2023. URL https://www.shodan.io/.

31 Tenable scan report formats, 2023a. URL https://docs.tenable.com/nessus/Content/ScanReportFormats.htm.

32 Tenable-sc, 2023b. URL https://www.tenable.com/products/tenable-sc.

33 Zmap, 2023. URL https://zmap.io/.

34 Omar Almazrouei and Pritheega Magalingam. The Internet of Things (IoT) network penetration testing model using attack graph analysis. In *2022 International Symposium on Multidisciplinary Studies and Innovative Technologies (ISMSIT)*, Oct 2022.

35 Rae Baker. *Deep dive exploring the real-world value of Open Source Intelligence.* Wiley, 2023.

36 Esteban Borges. Top 5 most popular port scanners in cybersecurity, Sep 2021. URL https://securitytrails.com/blog/best-port-scanners.

37 Richard L Craft. Red teaming in the age of IoT: thoughts on framing the next generation of technical vulnerability assessment. In *2017 12th System of Systems Engineering Conference (SoSE)*, 2017.

38 Jacob Fox. Offensive security within the software development life cycle (SDLC), Nov 2023. URL https://www.cobalt.io/blog/offensive-security-within-the-software-development-life-cycle.

39 Chad Maurice, Jeremy Thompson, and William Copeland. *The foundations of threat hunting: organize and Design Effective Cyber threat hunts to meet business needs*. Packt Publishing Ltd., 2022.

40 The Mentor. The conscience of a hacker, Jan 1986. URL http://phrack.org/issues/7/3.html.

41 Carnegie Mellon University. Social engineering, 2023. URL https://www.cmu.edu/iso/aware/dont-take-the-bait/social-engineering.html.

42 Micah Zenko. *Red team: how to succeed by thinking like the enemy*. Brilliance Audio, 2016.

3

Reinforcement Learning: Theory and Application

3.1 An Introduction to Reinforcement Learning (RL)

Many real-world problems for digital systems – driving, balancing, playing a video game – can be modeled as a sequence of decision-making steps. Each of these problems usually has an associated end-state, or objective such as arriving at a destination or successfully staying upright on a moving platform. Traditional methods of machine learning (ML) involve making predictions or assessments at a certain point in time such as finding the location of a dog in an image or predicting the probability of a sequence of authentication movements being lateral movement. Such paradigms, while effective, would not work well in solving the aforementioned sequential decision-making problems. Learning to make such decisions with the objective of reaching a given end-state will require a learning paradigm embodied in RL. RL and its associated concepts definition can best be illustrated by an example. Let us begin by setting up a hypothetical experiment wherein we are challenged with balancing a pencil upright on the tip of our finger. The challenge has four rules:

1. You can only use one hand to balance the pencil, and you can only do it by moving the finger on which the pencil rests left and right.
2. The challenge is to balance the pencil upright for as long as you can. For every second you balance it successfully, you earn 1 dollar.
3. If the pencil tips over, or falls, the experiment ends and your earnings stop at the time at which it ends.
4. The challenge can last up to a maximum of 20 s. If you reach 20 s, you earn 20 additional dollars.

Ostensibly, we begin the experiment by trying to develop a heuristic understanding of the pencil and its properties as you try to balance it: its sway as you move

Reinforcement Learning for Cyber Operations: Applications of Artificial Intelligence for Penetration Testing, First Edition. Abdul Rahman, Christopher Redino, Dhruv Nandakumar, Tyler Cody, Sachin Shetty, and Dan Radke.

your hand, the threshold at which it tips over, etc. We do this multiple times until the experiment fails. However, we take valuable pieces of learned information between each trial, which ultimately allows us to balance the pencil for longer and longer periods of time, maybe even reaching the 20-s mark. The longer we hold the pencil, the more money we earn, up to a maximum of 40 dollars. This is the essence of reinforcement learning (RL).

In the above example, we are the "agent"; the entity balancing the pencil. We take "actions," i.e., moving our hand left or right, to influence our "environment" (everything but us, the agent), that influence the pencil's "state" – defined here as the pencil's angular velocity. The objective of each experiment, or episode, is to hold the pencil upright for 20-time steps or, mathematically, vary the angular velocity of the pencil such that its angle relative to our finger does not cross the threshold beyond which the pencil tips. For every second, we hold the pencil up, we earn a "reward," a single dollar. Completing our objective gives us a larger reward, 20 dollars. Intuitively, the above process boils down to us taking steps that maximize our potential reward by keeping the pencil up for as long as we can. Realistically, we do a lot more than this when balancing a pencil, but those problems have essentially the same formulation.

All RL problems can be formulated as follows. An agent A interacting with an environment E by taking actions a to modify states s such that the expected reward R is maximized. The state here can be simply defined as the properties of the environment at any given moment. The logic by which the agent, A, chooses an action a given a state s is called a *policy*. Formally, a policy maps state spaces to action spaces, and every action will affect the environment E and its state s. Every *(state, action)* cycle is called a *time step*. At every time step, the agent earns a *reward*, which is larger for favorable actions. It should be noted here that the "quality" of an action is domain-dependent. For example, in the case of holding our pencil upright, a good action would be one in which we move our hand toward the direction in which the pencil is leaning to prevent its fall. The *objective* of an agent is to maximize the reward. An agent being rewarded more for good actions incentivizes the agent to select them more, *reinforcing* the agent's behavior (Figure 3.1).

Programmatically, an agent A interacts with its environment E using discrete messages in the form of a tuple of (state, action, and reward) as follows: (s_t, a_t, r_t) at time step t. Each of these tuples is called an *experience*. Although the agent could collect these experiences infinitely, that is not practically possible or even desirable. Consequently, we introduce a natural stopping point for the interaction by either introducing a terminal state or a maximum number of time steps where $t = T$. When the termination criteria are achieved, the agent stops collecting experiences; we call that collection of experiences an *episode*. RL agents generally learn from experience slower than humans do, often requiring hundreds or even millions of episodes to learn a good policy.

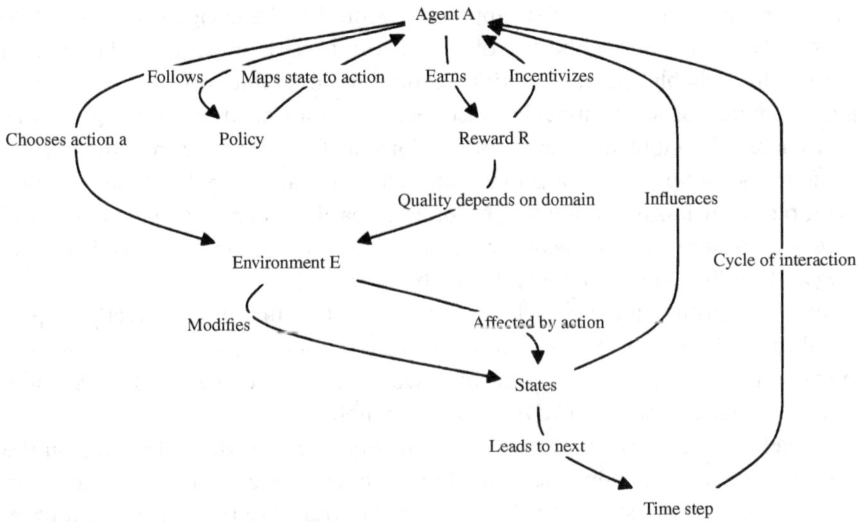

Figure 3.1 RL agent interaction with environment.

3.2 RL and Markov Decision Processes

In the real world, our environment is governed by physical phenomena that define how objects interact with each other and predicate how the environment transitions from one state to another. In the example of balancing a pencil on your fingertip, for instance, the physical laws of motion and gravity define how the pencil transitions between states as you take action. Similarly, in RL, we need to define a *transition function* that outlines how a simulated environment transitions between *states* given a series of *actions*. Let us explore the creation of a transition function by example.

Imagine that you are a taxi driver in a bustling city. Your main daily goal is to pick up as many passengers as possible to earn a good day's wage. To do this, you must decide where to drive to find these passengers. Should you go to the busy shopping district? The residential area? Or perhaps hang around the train station? Formally, you are trying to define a transition function that defines your next state, i.e., where you will find your next passenger. Now, let's say you tried to base this decision on every single place you've ever picked up a passenger before, right from your very first day on the job. That means, whenever you want to find your next passenger, you mentally sift through years of pick-up locations, times, weather conditions, and more. The transition function can be given by the equation:

$$s_{t+1} \sim P(s_{t+1} | (s_0, a_0), (s_1, a_1)...(s_t, a_t)) \tag{3.1}$$

where, at each time step t, the next state s_{t+1} is sampled from a probability distribution conditioned on the entire history. Simply put, your next state depends on all your previous states. This approach would be like navigating a city while constantly reading a map the size of a billboard. It is too much information, and you would probably spend more time reading the map than driving! Similarly, it is challenging for an RL model to also learn from a transition function such as this. There are simply too many combinations of effects that occurred in the past, particularly if an episode spans many time steps. Creating a policy based on such a complex transition function also becomes infeasible, given that an agent would have to consider its entire history when deciding on an action. This could take too long and could lead to instability during training [11].

However, given that the complexity of the transition function is directly proportional to the length of the conditioned history, it stands to reason that reducing the amount of history taken into account would improve efficiency. This is where Markov decision processes (MDPs) [12] are beneficial.

Instead of keeping track of your entire history of pick-ups, MDPs focus on the present moment, the "here and now." In your case, as a taxi driver, the "here and now" is your current state. This includes your current location, the current time, and the present weather conditions. For example, you might find yourself near a popular restaurant at 7 pm when it's raining. As a taxi driver using the principles of an MDP, your current state decides where to go next. If you're near the restaurant at dinner time, and it is raining (meaning fewer people are likely to walk), you might linger there. On a sunny weekday morning, when you're near a residential area, you might decide to wait for people commuting to their offices.

This method of decision-making allows you to make quick, informed decisions that feel intuitive and logical. Instead of getting overwhelmed with a sea of past data, you can focus on the most relevant information and adjust your actions according to the current situation. This not only simplifies the decision-making process but also makes it more flexible and responsive to changing conditions. Formally, this line of thinking defines the *Markov Property*, which outlines that a transition to the next state s_{t+1} depends only on the current state s_t and action a_t pair. The following equation represents this far simpler transition function:

$$s_{t+1} \sim P(s_{t+1} | (s_t, a_t)) \tag{3.2}$$

where the next state s_{t+1} can be sampled from a probability distribution conditioned only on the previous state-action pair. Despite the apparent simplicity of this idea, it is quite powerful in many computer applications involving robotics, simulation, and much more. It is important to note, though, that the environment must be constructed such that the previous state contains all necessary information, such as the weather and time of day in our example, required to make an informed decision.

An MDP is structured around four main components: states, actions, transition probabilities, and rewards.

- A set of **states (S)** are simply descriptions of the situation at hand. As a taxi driver, a state could be your current location, the time of day, and the weather. Each of these factors could impact where your next passenger might be.
- A set of **actions (A)** are the choices you can make in each state. As a taxi driver, your actions might be to drive toward the shopping district, wait by the train station, or head toward residential areas.
- **Transition probabilities ($P(s_{t+1}|(s_t, a_t))$)** tell you how likely it is that a particular action in a specific state will lead to a certain new state. For instance, if you're near the shopping district at 3 pm on a sunny day (your current state), and you choose to drive toward the residential area (your action), the transition probability might be the likelihood that you'll find a passenger there at this time under these conditions. Transition probabilities add a layer of uncertainty, or stochasticity, to the decision-making process, representing the fact that you cannot always predict exactly what will happen next.
- **Rewards ($R(s_t, a_t, s_{t+1})$)** are the benefits you get from making certain decisions. As a taxi driver, your reward might be the fare you collect from a passenger. In MDPs, you're usually trying to find the sequence of actions that will maximize your total reward over time.

One important assumption to note in RL is that agents do not have access to the transition function or reward functions directly as that would defeat the purpose of training. The MDP is simply used to define the behavior of the environment. RL agents must learn to approximate the respective functions by repeatedly interacting with the environment and learning from the episode 3-tuple: (s_t, a_t, r_t).

Our picture of an MDP is almost complete. To fully understand the mathematical objective of an MDP, we need to introduce the concept of an objective function, sometimes also referred to as a loss or cost function. The objective function is essentially the goal you are trying to achieve, expressed mathematically. For the taxi driver, the objective is to maximize the total fare collected over a shift. The mathematical objective function represents this goal.

In an MDP, the objective function often involves the sum of rewards collected over time, where each reward is the result of a particular action taken in a certain state. In the case of our taxi driver, each reward might be the fare from a single passenger. The goal of an *agent* over an *episode* is to maximize the expected reward or maximize the objective function. Following that logic, the objective function could be given as the return $R(\tau)$ over a trajectory of experiences $\tau = (s_0, a_0, r_0)...(s_T, a_T, r_T)$ using the equation:

$$R(\tau) = \sum_{t=0}^{t=T} (r_t) \tag{3.3}$$

However, we need to consider the fact that not all rewards are created equal. Specifically, rewards that come later are usually considered less valuable than immediate rewards. This reflects the principle of delayed gratification: a bird in the hand is worth two in the proverbial bush. To accomplish this, we introduce a discount factor $\gamma \in [0, 1]$ which diminishes the value of future rewards. We can modify the above equation to include a discount factor as:

$$R(\tau) = \sum_{t=0}^{t=T} (r_t * \gamma^t) \tag{3.4}$$

Here, we see that as the time t gets larger, the value of the reward would get lower. The objective function $J(\tau)$ is simply the expected reward over many episodes or trajectories.

$$J(\tau) = E_{\tau \sim \pi}[R(\tau)] \tag{3.5}$$

The objective function is an average or expected value of cumulative reward over many episodes. Recall that the MDP or environment is defined by transition probabilities, which add inherent uncertainty. The average or expected value is used here to account for the MDP's stochasticity.

The discount factor $\gamma \in [0, 1]$ controls how "nearsighted" or "farsighted" the agent is during training. If we set $\gamma = 1$, rewards from all time steps are weighted equally, giving us the reward equation we started with. Conversely, if $\gamma = 0$, the agent will be completely shortsighted, paying attention to only the most immediate reward, i.e., $R(\tau) = r_t$. γ is an important hyperparameter we modify (or lever we pull) when training RL agents as problems often become noticeably harder or easier to solve based on the choice of discount factor. This completes our definition of an MDP.

3.3 Learnable Functions for Agents

Now that we have built a strong foundation of MDPs, we can explore how agents learn from in an environment. We have already alluded to the idea of agents approximating an optimal policy that maps states to good actions but that is just one aspect of an environment an agent can learn. Realistically, an agent can also learn other facets of environments, such as the values of states or environmental configurations themselves, which could be valuable. We will explore all three primary functions, namely:

1. The Policy Model
2. The Value-Based Model
3. Model-Based Learning

3.3.1 The Policy Model

In the context of our taxi drivers, imagine that they have a guidebook that tells them what to do in every possible situation or state. For instance, the guidebook might tell them that at 7 pm on a rainy day, they should head to the shopping district because that's where they will likely find the most passengers. This guidebook is what we refer to as a policy in RL.

In policy learning, the objective is to learn the best guidebook or policy directly. That is, the objective is to the optimal mapping from states to actions such that the expected reward function is maximized. A policy, π, can be stochastic and select appropriate actions for a given state by sampling them from a probability distribution, given by $\pi(a|s)$. The policy, π, for a given state, can be given by the equation

$$\pi = \text{argmax}_a[R(s,a) + \gamma \sum_{s'} P(s'|(s,a))V_\pi(s')] \tag{3.6}$$

known as the Bellman Optimality Equation [6]. Here, the equation expresses the policy as the action that maximizes the current reward plus the discounted value of all future states, giving us a recursive method by which to compute a policy (e.g., depth-first search). Note here that for brevity's sake, it is customary to write successive actions, states, or rewards using "prime" notation, s and s'.

The algorithm tries different actions in different states and gradually learns which actions lead to the best outcomes, updating the guidebook over time. The advantage of policy-based methods is that they can handle situations where the action space is large or even continuous and can learn to balance between exploration (trying new actions) and exploitation (sticking to known good actions). However, they tend to be computationally expensive to evaluate and often converge to local optima as opposed to global optima. REINFORCE is an example of a popular policy-learning model [13].

3.3.2 The Value-Based Model

Now, suppose instead of a guidebook, the taxi driver had a rating system that assigned a score to each possible state, indicating how good that state is in terms of potential earnings. For instance, being near a concert venue just as a concert ends might have a high score because many people will be looking for taxis. This rating system is what we refer to as a value function in RL.

Value functions in RL generally come in two different flavors: $V^\pi(s)$ and $Q^\pi(s,a)$. They are defined as:

- $V^\pi(s)$ is a state-value function defined purely as the expected value of a given state s in general, or across all actions. Suppose that at different times, our taxi driver finds themselves at the city's train station. The state-value function, in this case, could be represented as V(train station), which would be the expected total fare the driver can earn from picking up passengers at the train station, given their current policy. This value could be influenced by factors such as time of day, day of the week, and weather conditions, and it is learned based on the taxi driver's past experiences and the rewards received, that is, the fares they have collected in the past from that location. The taxi driver then uses these learned values to inform their future decisions, aiming to choose locations (states) with higher expected rewards.

- $Q^{\pi}(s, a)$, also called the Q-value function, denotes the value of a given state-action pair. Continuing with our taxi driver analogy, let's consider an instance where the driver is currently at the city's train station and is deciding between two actions: wait for passengers at the station or drive to the nearby shopping mall. In this case, the Q-value function could be represented as *Q*(train station, wait) and *Q*(train station, drive to mall). These Q-values represent the expected total fare the driver can earn from either waiting at the train station or driving to the shopping mall, given their current policy. As before, these values are influenced by various factors such as time of day, day of the week, and weather conditions, and they are learned based on the taxi driver's past experiences and the rewards received.

Generally, Q-value functions are far more common when training RL models than state-value functions as they can be translated to policies most easily. This can be done by sampling the action in a given that produces the highest Q-value. State-Action-Reward-State-Action (SARSA), deep Q-networks, and deep Q-networks (DQN) with prioritized experience replay (ER) [14] are some common implementations of value-based models. The advantage of value-based methods is that they can be very efficient, especially when the state and action spaces are large. However, they are not guaranteed to converge to an optimum and are only applicable to discrete action spaces.

3.3.3 Model-Based Learning

Finally, consider that our taxi driver could have a model of the city and its inhabitants. This model would include information about how people move around the city at different times under different conditions. With this model, the taxi driver could predict what might happen if they drive to the shopping district at 7 pm on a rainy day. This model of the environment is what we refer to as a model in RL.

Algorithms of this nature build an internal representation of the environment in which they operate. Once such a model is constructed, the agents can use their internal understandings to predict outcomes a few time steps in the future without having to interact with the environment itself. An agent can predict, or "imagine," many different trajectories of states based on different action sequences and ultimately pick the most beneficial action to take at a given time step based on the predictions.

The key advantage to model-based methods is that they are incredibly sample efficient. Because they do not need to interact with the environment to predict or imagine a few time steps in advance, they often require much fewer experiences to learn than value-based or policy-based methods. However, well-defined environments without too much stochasticity are hard to come by. Model-based methods also tend to underperform in environments that have a large number of states and action spaces simply because there are too many possibilities to simulate [4]. However, if such hurdles are overcome, model-based methods are often much more efficient than other methods while maintaining comparable performance. Monte Carlo tree search (MCTS) [1], linear quadratic regulators (iLQR), and model predictive control (MPC) [5] are examples of model-based algorithms.

3.3.4 Combining Methods

By now, it is clear that all three of the above methods have their individual benefits and drawbacks. One method to enhance performance of RL methods could be to use combinations of the three models above to create a chimera approach that embodies the strengths of each approach.

Let us consider a combination of policy-based and value-based methods wherein a modeling approach learns both a policy and a value function in a given environment. These types of approaches are called *Actor-Critic* approaches, where one agent acts, or performs actions based on a learned policy, and another agent critiques or provides feedback on actions based on a learned value functions for a state-action pair.

So, in the case of our example with the taxi driver, the actor is responsible for making decisions – it's part of the model that determines how our taxi driver should behave given their current state. In other words, if our taxi driver is in the downtown district on a rainy evening, the actor is part of the model that decides what action the taxi driver should take – perhaps driving to a theater that's about to let out a show. The critic, on the other hand, is responsible for evaluating the actions taken by the actor. The critic does this by estimating the value of the state resulting from the action – basically, predicting the total future reward that the taxi driver will receive if they follow the current policy. It then provides feedback to the actor, helping it adjust its decisions to increase the expected rewards.

The key benefit of such an approach is that the actor can learn a better policy using value functions which are often more informative than just a series of rewards for actions. There have been many interesting approaches to actor-critic algorithms, some of which will be the key focus of this book: trust region policy optimization (TRPO), proximal policy optimization (PPO), deep deterministic policy gradients (DDPG), and others [7, 10]. Other methods also exist that combine model-based methods with others, but they will not be covered in this work.

3.4 Enter Deep Learning

An autonomous vehicle operates in a complex and continuously changing environment, needing to make a series of decisions such as when to accelerate, when to brake, and when to turn. The car's objective is to navigate to a destination safely, efficiently, and comfortably for its passengers.

At its heart, the task of an autonomous vehicle is a RL problem. The car, acting as an agent, interacts with its environment (the road, other vehicles, pedestrians, etc.), taking actions based on its current state (location, speed, nearby obstacles, etc.), and receiving rewards (positive for reaching the destination quickly and safely, negative for dangerous situations or violating traffic rules).

Deep learning enters the scene as a powerful tool to handle the high-dimensional inputs and complex function approximations often required in such RL problems.

- **Perception:** Autonomous vehicles use sensors like cameras, LIDAR, and radar to gather data about their environment. These data are high-dimensional and unstructured. Deep learning is an effective tool for processing and interpreting this kind of data. For example, convolutional neural networks (CNNs) can process images from cameras to identify other vehicles, pedestrians, and road signs.
- **Policy Learning:** The decision-making process, or policy, of an autonomous vehicle, can be complex and needs to handle a wide variety of scenarios. Deep neural networks can be used to approximate this policy function, effectively learning from data to make decisions. This approach, known as deep reinforcement learning (DRL), combines the decision-making framework of RL with the function approximation capabilities of deep learning.
- **Value Estimation:** In many RL methods, it's necessary to estimate the expected future reward, or "value," of different states or actions. This can be a challenging task when the state and action spaces are large or continuous. Again, deep learning comes to the rescue, with deep neural networks used to approximate these value functions.

Deep learning models are the most powerful function approximation tools we know of and use today. The integration of deep learning into RL leads into a paradigm we call DRL. However, integrating DL into RL poses some challenges and nuances. As such, it is first beneficial to understand how DL models learn. Deep learning models are essentially composed of interconnected layers of artificial neurons, or nodes, that can learn to represent complex patterns in data. These models are trained using a process called backpropagation and an optimization technique, such as gradient descent.

A simplified version of the process is presented as follows:

- **Forward Pass:** The model takes input data and makes predictions. The data is passed through each model layer, with each layer applying learned transformations to the data and passing it on to the next layer. At the end of this process, the model produces a prediction.

- **Loss Calculation:** The model's prediction is compared to the true value, and the difference is quantified into a single number, known as the loss. The loss function measures how well the model performs, with lower values indicating better performance.
- **Backward Pass (Backpropagation):** The model then works backward, calculating how much each node in each layer contributed to the final loss. This involves computing the derivative of the loss function concerning the weights of the model, indicating how much a small change in each weight would affect the loss.
- **Update Weights:** The model weights are then updated to minimize the loss. Typically, an optimization algorithm like gradient descent is used. The algorithm adjusts each weight in the direction that reduces the loss.
- **Repeat:** The process is repeated with new input data until the model's predictions are satisfactory or the model stops improving. When integrating RL into this process, there are a few key differences (Figure 3.2).

Firstly, in RL, there is often no explicit "correct answer" for the model to learn from, as in supervised learning. Instead, the model learns from the rewards and punishments it receives as it interacts with its environment. Secondly, the loss function in an RL setting often involves estimating future rewards, which introduces additional uncertainty and complexity to the learning process. Finally, because the model's actions can change the data it receives, RL involves a degree of exploration – trying different actions to discover which ones lead to the highest rewards. This contrasts the static datasets typically used in supervised deep learning, and it introduces additional challenges in balancing exploration (trying

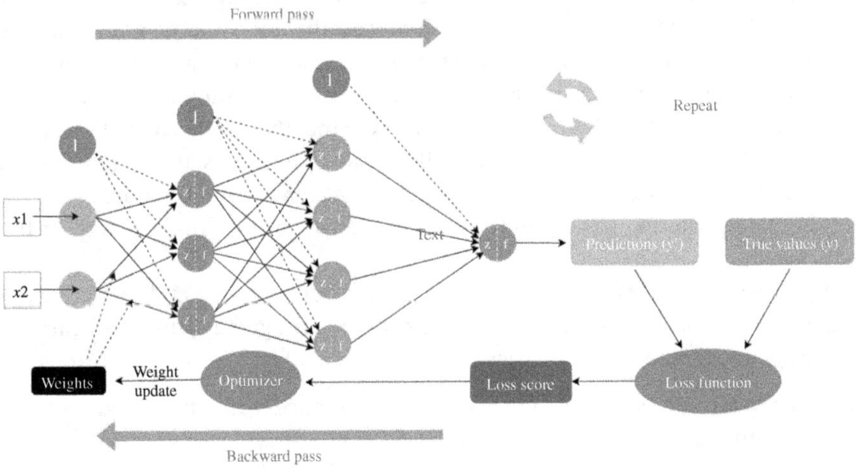

Figure 3.2 The process of training a deep learning model.

new things) with exploitation (sticking with what works). Lastly, collecting experiences in DRL is a time-consuming process as the agent must go through each episode, wait for the environment to respond, and move forward.

Despite these challenges, combining deep learning with RL has shown remarkable success in many areas, including gameplay, robotics, and autonomous vehicles. Deep learning's ability to handle high-dimensional inputs and learn complex patterns, combined with the decision-making framework of RL, makes for a powerful combination.

3.5 Q-Learning and Deep Q-Learning

Q-Learning is one of the most widely used and successful RL algorithms. At its core, Q-Learning aims to learn an action-value function, denoted $Q(s, a)$, that gives the expected future reward for taking a given action in a given state s. Intuitively, we can think of the Q-function as a lookup table that the agent consults to decide which action to take in any situation. The table stores running estimates of the quality of each action in each state, based on the rewards the agent has received so far. The agent exploits this knowledge to select actions leading to high rewards. More formally, Q-learning iterates through a process of observing the current state s, choosing an action a, receiving a reward r, observing the next state s', and updating its estimated Q-values as follows:

$$Q(s,a) < -Q(s,a) + \alpha * (r + \gamma * \max{}_a' Q(s', a') - Q(s,a))$$

Where α is the learning rate and γ is the discount factor for future rewards. This update rule shifts the estimate for $Q(s, a)$ toward the observed reward r plus the maximum future reward given the next state s', discounted by gamma. By repeating this process over many episodes of experience, Q-learning converges to the optimal action values. It is important to note here that the Bellman equation above discounts the *maximum* Q value over all actions of a given state rather than an immediate next state. This implies that no part of the Q-value function depends on the data-gathering policy, and is an off-policy algorithm independent of the policy being followed.

A policy can then be derived from the learned Q-function by simply selecting the highest valued action in each state. This is known as the greedy policy. Q-Learning has attractive theoretical properties:

- It is model-free, learning directly from experience without a model of the environment.
- It is off-policy, able to learn about the optimal policy independently of the agent's actions.

- It converges to the optimal policy given sufficient exploration.
- It is computationally lightweight, involving simple value updates.

These strengths have led to the widespread adoption of Q-Learning across many domains. However, vanilla Q-Learning relies on tabular storage of Q-values, limiting its applicability in problems with large state and action spaces. This is where Deep Q-Learning [3] comes in. It combines the strengths of Q-Learning with the representation power of deep neural networks. Rather than representing $Q(s, a)$ via a table, a neural network is used to approximate the Q-function. The network takes the state s as input and outputs Q-value estimates for each possible action a. It can be trained via applying the same Q-learning updates, using backpropagation to adjust the network weights to minimize the loss between the predicted and target Q-values. Over time, the neural network learns to approximate the optimal action-value function. The agent can then select actions greedily based on the network's outputs. The key advantage is that deep neural networks can handle higher dimensional state inputs and learn complex non-linear relationships. This enables deep Q-learning to scale to problems with large state and action spaces.

3.5.1 Boltzmann Policies and Experience Replay

In the field of RL, the dilemma of exploration versus exploitation plays a pivotal role. Exploration involves trying new actions to gain more knowledge about the environment, while exploitation uses the current knowledge to make decisions that seem best. Striking a balance between these two strategies is crucial for the effective training of RL agents.

One of the simplest methods to address this challenge is the ϵ-greedy policy. This strategy, with a certain probability ϵ, encourages the agent to explore by choosing actions randomly. Conversely, with probability $1 - \epsilon$, it exploits the agent's existing knowledge by selecting the best-known action. However, the ϵ-greedy approach, with its binary decision-making and static exploration rate, lacks contextual sensitivity and adaptability. It does not consider the relative values of different actions, treating them all equally during exploration.

In contrast, Boltzmann policies, also known as softmax action selection, offer a more nuanced approach. These policies assign probabilities to each action based on their Q-values, using a softmax function. The probability of choosing a specific action in a state is given by:

$$P(a|s) = \frac{e^{Q(s,a)/\tau}}{\sum_{a'} e^{Q(s,a')/\tau}} \tag{3.7}$$

Here, $P(a|s)$ is the probability of selecting action a in state s, and τ is a temperature parameter that modulates the level of exploration. A high value of τ leads to more uniform probabilities, encouraging exploration, while a low τ favors exploitation by making the policy more sensitive to higher Q-values. Practically, Boltzmann policies use a higher value for τ at the start of training to encourage exploration and gradually decay it over time to exploit what has been learned. However, care must be taken not to decay τ too quickly as that may result in the agent converging to a local optimum.

The advantages of Boltzmann policies are manifold. They dynamically adapt the exploration level based on the agent's confidence in its action values, offer context-aware decision-making by considering the relative merits of different actions, and provide a balanced approach to transitioning between exploration and exploitation.

Furthermore, a key advantage DQN as an off-policy algorithm is that it is not temporally dependent on experience sampling like on-policy algorithms are. This means that one can improve the computational efficiency of training a DQN agent by storing prior experiences for future training. This was first observed by Long-Ji Lin [2] who called this *ER*. ER memory stores the n most recent experience an agent has gathered as a queue. Consequently, as new experiences are gathered, older experiences are removed to make room. During training, one or more batches of experiences are sampled from the ER memory for use in the DQN algorithm. ER memory should generally be large enough to contain many experiences, and typically are large enough to store up to 1,000,000 experiences. ER helps stabilize training by reducing the variance of parameter updates.

3.5.2 Implementing DQN

To illustrate an end-to-end implementation of DQN, let's walk through a Python script using the PyTorch framework, set in the context of the CartPole environment from OpenAI's Gym.

3.5.2.1 The CartPole Environment

The CartPole is a benchmark environment in RL. The agent's task is to balance a pole on a cart that moves horizontally. Each time the pole remains upright, the agent receives a reward. The challenge lies in keeping the pole balanced by moving the cart, aiming to maximize the cumulative reward (Figure 3.3).

3.5.2.2 DQN Architecture

The initial step in DQN implementation involves defining the Q-network. This neural network approximates the Q-values, which are the expected rewards for executing actions in particular states. Our DQN is composed of two fully connected layers, an architecture well-suited for the CartPole environment (Figure 3.4).

```
import gym
import torch
import torch.nn as nn
import torch.optim as optim
import numpy as np
import random
from collections import deque

# Initialize environment and parameters
env = gym.make('CartPole-v1')
state_dim = env.observation_space.shape[0]
action_dim = env.action_space.n
```

Figure 3.3 Python code for initializing the environment and parameters in CartPole-v1 using PyTorch and OpenAI Gym.

```
# Define the Q-Network
class DQN(nn.Module):
    def __init__(self, state_dim, action_dim):
        super(DQN, self).__init__()
        # Define a simple neural network with two fully connected
                                              layers
        self.fc1 = nn.Linear(state_dim, 64)
        self.fc2 = nn.Linear(64, action_dim)
    def forward(self, x):
        x = torch.relu(self.fc1(x))
        x = self.fc2(x)
        return x
```

Figure 3.4 Definition of the DQN class and network architecture.

3.5.2.3 Boltzmann Policy for Action Selection

A crucial element of DQN is the strategy for action selection. We utilize a Boltzmann policy, a method that probabilistically chooses actions based on their Q-values. This policy strikes a balance between exploration and exploitation, dynamically adjusting the probabilities of actions relative to their value (Figure 3.5).

3.5.2.4 Experience Replay Mechanism

To improve learning efficiency and stability, we incorporate an ER mechanism. The agent's interactions with the environment, encompassing states, actions, rewards, and subsequent states, are stored in a replay buffer. This buffer allows the agent to learn from previous experiences, reducing correlations between consecutive updates and enhancing data use (Figure 3.6).

```
def boltzmann_policy(q_values, tau=1.0):
    preferences = q_values / tau
    max_preference = torch.max(preferences)
    exp_preferences = torch.exp(preferences - max_preference)
    probs = exp_preferences / torch.sum(exp_preferences)
    action = torch.multinomial(probs, 1).item()
    return action
```

Figure 3.5 Python function implementing a Boltzmann policy.

```
# Replay Buffer
class ReplayBuffer:
    def __init__(self, capacity):
        self.buffer = deque(maxlen=capacity)
    def push(self, state, action, reward, next_state, done):
        self.buffer.append((state, action, reward, next_state,
                                          done))
    def sample(self, batch_size):
        state, action, reward, next_state, done = zip(*random.
                                          sample(self.buffer,
                                          batch_size))
        return state, action, reward, next_state, done
    def __len__(self):
        return len(self.buffer)
```

Figure 3.6 Python class implementing a replay buffer.

3.5.2.5 The Training Process

The core of the DQN implementation is the training loop. Over a predefined number of episodes, the agent interacts with the environment in steps, where it:

1. Observes the current state.
2. Chooses an action based on the Boltzmann policy.
3. Executes the action and notes the reward and the next state.
4. Stores this experience in the replay buffer.

Upon gathering enough data, the agent randomly samples batches from the buffer for training. This process involves:

- Computing current Q-values from the DQN for the sampled states and actions.
- Calculating target Q-values using rewards and the maximum Q-values for the next states, adjusted by a discount factor.
- Minimizing the mean squared error between the current and target Q-values to update the DQN.

```
dqn = DQN(state_dim, action_dim)
optimizer = optim.Adam(dqn.parameters(), lr=0.001)
# Replay buffer parameters
replay_buffer = ReplayBuffer(10000)
batch_size = 64
replay_start_size = 1000  # Number of experiences to collect before
                                  starting training
# Training parameters
num_episodes = 500
max_steps_per_episode = 100
tau = 1.0  # Temperature parameter for Boltzmann policy
gamma = 0.99  # Discount factor for future rewards
# Main training loop
for episode in range(num_episodes):
    state = env.reset()
    total_reward = 0
    for step in range(max_steps_per_episode):
        state_tensor = torch.FloatTensor(state).unsqueeze(0)
        with torch.no_grad():
            q_values = dqn(state_tensor)
        action = boltzmann_policy(q_values, tau)
        next_state, reward, done, _ = env.step(action)
        total_reward += reward
        replay_buffer.push(state, action, reward, next_state, done)
        state = next_state
        # Check if replay buffer has enough samples to start
                                        learning
        if len(replay_buffer) > replay_start_size:
            sampled_states, sampled_actions, sampled_rewards,
                                        sampled_next_states,
                                        sampled_dones =
                                        replay_buffer.sample(
                                        batch_size)
            # Convert to tensors
            states = torch.FloatTensor(sampled_states)
            actions = torch.LongTensor(sampled_actions)
            rewards = torch.FloatTensor(sampled_rewards)
            next_states = torch.FloatTensor(sampled_next_states)
            dones = torch.FloatTensor(sampled_dones)
            # Get current Q values
            q_values_current = dqn(states).gather(1, actions.
                                        unsqueeze(1)).squeeze
                                        (1)
            # Compute target Q values
            next_state_values = dqn(next_states).max(1)[0]
            target_q_values = rewards + gamma * next_state_values *
                                        (1 - dones)
            # Calculate loss
            loss = nn.MSELoss()(q_values_current, target_q_values)
            # Optimize the model
            optimizer.zero_grad()
            loss.backward()
            optimizer.step()
        if done:
            break
```

```
torch.save(dqn.state_dict(), "dqn_model.pth")
```

Figure 3.7 Saving a PyTorch model.

3.5.2.6 Post-Training

Post-training, the learned policy is encapsulated in the saved model parameters. This comprehensive implementation of DQN in the CartPole environment showcases the effective fusion of deep learning with Q-learning to address complex decision-making tasks. Incorporating a Boltzmann policy and ER mechanism marks a significant enhancement to the basic DQN, leading to a more robust and efficacious learning process (Figure 3.7).

3.6 Advantage Actor-Critic (A2C)

After exploring DQN, we transition to advantage actor-critic (A2C), a pivotal algorithm in modern RL that refines the ideas introduced in DQN. A2C represents a sophisticated blend of value-based and policy-based methods, employing a dual-structure approach with an "actor" for policy execution and a "critic" for evaluating the chosen actions. Unlike DQN, which focuses on learning a value function, A2C distinctly separates the policy (actor) and the value estimation (critic) roles. This separation allows for more precise and efficient learning processes, particularly beneficial in environments with complex or continuous action spaces. Formally, the actor learns a parameterized policy function and the critic learns a value function to evaluate state-action pairs. The "advantage" in A2C is key to understanding its efficiency. It refers to the benefit of taking a specific action in a given state, as compared to the average value of all possible actions in that state. By calculating this advantage, A2C enables the actor to make informed decisions that are relative not just to the state but also to the action being evaluated. This approach contrasts with DQN, where policy decisions are derived indirectly from the value function. In A2C, the policy is learned directly, leading to more refined action choices and potentially faster convergence. The critic's role is to provide a baseline against which the advantage of actions taken by the actor can be measured, thereby reducing the variance in policy updates and contributing to more stable training. Learning a value function with a critic can also be more informative than using rewards from the environment directly when approximating a policy. This can be seen primarily in the case of an environment with sparse rewards, where a learned value function could provide a denser reinforcing signal than the reward in and of itself.

The upcoming sections will delve into the technical workings of A2C, illustrating how it effectively combines the actor's direct policy learning with the critic's evaluative feedback to tackle complex challenges in RL.

3.6.1 The Actor

The actor learns a parameterized policy, π_θ, that predicts an action given a state in the environment. In policy gradient methods, such as A2C, the policy, represented as $\pi_\theta(a|s)$ where θ are the policy parameters and a and s represent the action and state, is typically modeled using a neural network. The policy gradient formula is given by:

$$\nabla_\theta J(\theta) = \mathbb{E}_{\pi_\theta}\left[\nabla_\theta \log \pi_\theta(a|s) \cdot A(s,a)\right] \tag{3.8}$$

Where:

- $\nabla_\theta J(\theta)$ is the gradient of the performance measure J with respect to the policy parameters θ.
- $\mathbb{E}_{\pi_\theta}$ denotes the expectation under the policy π_θ.
- $\nabla_\theta \log \pi_\theta(a|s)$ is the gradient of the logarithm of the policy, indicating the likelihood of taking action a in state s under the policy π_θ.
- $A(s,a)$ represents the advantage function, quantifying the relative benefit of taking action a in state s over the average action.

3.6.2 The Critic and Advantage

The critic is responsible for ascribing value to a given state-action pair. In A2C, this is performed by computing the "advantage"; the value of a state-action pair relative to all other actions in that state. Intuitively, that translates to the equation $A^\pi(s_t,a_t) = Q^\pi(s_t,a_t) - V^\pi(s_t)$, or plainly, the difference between the Q-function and Value function for a given state-action pair. The advantage is a relative measure that normalizes reward for a given state. Furthermore, the advantage only affects the value of a future trajectory and changes in the value in future time steps.

The value of the advantage is usually derived and estimated from V^π. The process involves learning a parameterized value function, V^π, from which Q^π can be estimated. Calculating the value function, V^π, which represents the expected return from a state s, is generally less complex than computing the Q function, Q^π, for every action a in every state s. This efficiency is particularly significant in environments with large or continuous action spaces. The following equation describes the generalized advantage estimation (GAE) formula [8], that computes the advantage for a given state-action pair.

$$\hat{A}_t^{\mathrm{GAE}(\gamma,\lambda)} = \sum_{l=0}^{\infty}(\gamma\lambda)^l \delta_{t+l}^V \tag{3.9}$$

Where:

- $\hat{A}_t^{\mathrm{GAE}(\gamma,\lambda)}$ is the estimated advantage at time t.
- γ is the discount factor for future rewards.
- λ is a parameter that balances bias and variance in the estimate.
- δ_{t+l}^V is the temporal difference (TD) error at time $t+l$, defined as $r_{t+l} + \gamma V(s_{t+l+1}) - V(s_{t+l})$.

GAE averages multiple advantage function estimates, each with a different look-ahead window l, weighted by $(\gamma\lambda)^l$. This method allows for a flexible and efficient trade-off between bias and variance, and by adjusting λ, one can control the influence of future rewards on the advantage estimation. This leads to more stable and effective policy updates in A2C and similar algorithms.

3.6.3 Implementing A2C

We will now explore an example where we implement the A2C algorithm on PyTorch to train on the CartPole environment explored in Section 3.5.2 (Figure 3.8).

3.6.3.1 Actor and Critic Networks

The script begins by defining two neural network classes, actor and critic. The actor-network, responsible for policy execution, outputs a probability distribution over possible actions given a state. It uses a softmax layer to ensure the output probabilities sum to one. The critic network, on the other hand, estimates the value of a given state. Both networks are simple feed-forward models with one hidden layer (Figure 3.9).

3.6.3.2 The GAE Function

A key function in the script is "compute_gae," which calculates the GAE. GAE is a technique to estimate the advantage function, a measure of the relative value

```python
import gym
import torch
import torch.nn as nn
import torch.optim as optim
import numpy as np
# Initialize Gym environment
env = gym.make('CartPole-v1')
state_size = env.observation_space.shape[0]
action_size = env.action_space.n
```

Figure 3.8 A2C initial python imports.

```
    # Define the Actor network
class Actor(nn.Module):
    def __init__(self, state_size, action_size):
        super(Actor, self).__init__()
        # Simple feed-forward network with one hidden layer
        self.network = nn.Sequential(
            nn.Linear(state_size, 128),
            nn.ReLU(),
            nn.Linear(128, action_size),
            nn.Softmax(dim=-1)  # Softmax to output a probability
                                  distribution over
                                  actions
        )
    def forward(self, state):
        return self.network(state)
# Define the Critic network
class Critic(nn.Module):
    def __init__(self, state_size):
        super(Critic, self).__init__()
        # Similar network structure but outputs a single value (
                                value of a state)
        self.network = nn.Sequential(
            nn.Linear(state_size, 128),
            nn.ReLU(),
            nn.Linear(128, 1)
        )
    def forward(self, state):
        return self.network(state)
```

Figure 3.9 Python code implementing the actor and critic networks in A2C.

of actions compared to a baseline. It strikes a balance between bias and variance in the advantage estimates, improving the stability and efficiency of the learning process. GAE considers a series of TD errors, discounted and weighted, to provide a more stable estimate of advantages (Figure 3.10).

3.6.3.3 Training the Model

The script enters a training loop, where the agent interacts with the CartPole environment over a predefined number of episodes. In each episode, the agent collects states, rewards, and actions. After each action, it stores the log probability of the action, the estimated value from the critic, the received reward, and a mask indicating whether the episode is continuing.

Once an episode concludes, the script computes the returns using the "compute_gae" function and uses these to calculate the advantages. These advantages are then used to update both the actor and critic networks. The actor's loss is calculated to encourage actions that lead to higher advantages, effectively improving the

```
# Function to calculate Generalized Advantage Estimation (GAE)
def compute_gae(next_value, rewards, masks, values, gamma=0.99,
                                tau=0.95):
    values = values + [next_value]
    gae = 0
    returns = []
    for step in reversed(range(len(rewards))):
        delta = rewards[step] + gamma * values[step + 1] * masks
                                        [step] - values[step]
        gae = delta + gamma * tau * masks[step] * gae
        returns.insert(0, gae + values[step])
    return returns
```

Figure 3.10 Python function for generalized action estimation.

policy. The critic's loss is computed as the mean squared error of the advantages, aiming to improve the state-value estimation.

```
# Create actor and critic networks
actor = Actor(state_size, action_size)
critic = Critic(state_size)
# Set up optimizers for both networks
actor_optimizer = optim.Adam(actor.parameters(), lr=1e-3)
critic_optimizer = optim.Adam(critic.parameters(), lr=1e-3)
# Training parameters
num_episodes = 1000
# Main training loop
for episode in range(num_episodes):
    state = env.reset()
    done = False
    episode_reward = 0
    log_probs = []  # Store log probabilities of the actions
    values = []     # Store value estimates
    rewards = []    # Store rewards
    masks = []      # Store done masks (1 for not done, 0 for done)
    while not done:
        state_tensor = torch.FloatTensor(state)
        action_probs = actor(state_tensor)  # Get action
                                            probabilities from the
                                            actor
        value = critic(state_tensor)          # Get value estimate
                                            from the critic
        action = np.random.choice(np.arange(action_size), p=
                                            action_probs.detach().
                                            numpy())
        next_state, reward, done, _ = env.step(action)
        log_prob = torch.log(action_probs[action])
        episode_reward += reward
        log_probs.append(log_prob)
        values.append(value)
```

```
        rewards.append(reward)
        masks.append(1 - done)
        state = next_state
    next_state_tensor = torch.FloatTensor(next_state)
    next_value = critic(next_state_tensor)
    returns = compute_gae(next_value, rewards, masks, values)
    log_probs = torch.stack(log_probs)
    returns = torch.cat(returns).detach()
    values = torch.cat(values)
    advantage = returns - values
    # Calculate losses for both actor and critic
    actor_loss = -(log_probs * advantage.detach()).mean()
    critic_loss = advantage.pow(2).mean()
    # Perform backpropagation
    actor_optimizer.zero_grad()
    critic_optimizer.zero_grad()
    actor_loss.backward()
    critic_loss.backward()
    # Update both networks
    actor_optimizer.step()
    critic_optimizer.step()
    print(f'Episode {episode}, Total Reward: {episode_reward}')
```

3.7 Proximal Policy Optimization

While A2C represents a significant step forward in RL algorithms, it does come with notable challenges. A primary concern with A2C is its sample inefficiency. A2C typically requires a large number of interactions with the environment to effectively learn, which can be impractical or costly in real-world scenarios or complex simulations.

Another critical issue is inconsistency in policy updates. A2C updates can sometimes be overly aggressive, leading to significant policy shifts. This can cause instability in training, as the policy might oscillate or diverge, especially in the face of complex or high-dimensional decision spaces.

Lastly, A2C often exhibits high variance in reward signals. This variance can impede the learning process, as the algorithm struggles to discern between beneficial policy changes and stochastic environmental noise. All of the above factors lead to instability during training, performance collapse, and inefficient training processes. These challenges necessitate an approach that maintains steady and efficient learning, paving the way for the development of TRPO.

3.7.1 Trust Region Policy Optimization (TRPO)

TRPO [7] was developed to improve the stability and efficiency of policy gradient methods, focusing on consistent and reliable policy updates. Its core innovation is the introduction of a "trust region" to constrain policy updates. This ensures that new policies do not deviate excessively from old policies, maintaining stability in the learning process.

The objective function in TRPO includes a surrogate advantage function and is expressed as:

$$\max_{\theta} \mathbb{E}_{s,a\sim\pi_{\text{old}}}\left[\frac{\pi_\theta(a|s)}{\pi_{\text{old}}(a|s)}\hat{A}^{\pi_{\text{old}}}(s,a)\right] \tag{3.10}$$

where $\hat{A}^{\pi_{\text{old}}}(s,a)$ is the estimated advantage function under the old policy. The ratio $\frac{\pi_\theta(a|s)}{\pi_{\text{old}}(a|s)}$ reflects the divergence of the new policy from the old policy for given state-action pairs.

To ensure controlled policy updates, TRPO includes a constraint based on the Kullback-Leibler (KL) divergence:

$$\mathbb{E}_{s\sim\pi_{\text{old}}}\left[D_{\text{KL}}(\pi_{\text{old}}(\cdot|s)||\pi_\theta(\cdot|s))\right] \leq \delta \tag{3.11}$$

This constraint forms a "trust region" around the old policy, permitting exploration and improvement within the bounds of stability.

Despite its advancements, TRPO introduced computational complexities, especially in enforcing the KL constraint. This complexity, particularly in high-dimensional spaces or with deep neural networks, set the stage for the development of PPO, which aims to simplify these computations while retaining the core benefits of stable and reliable policy updates.

3.7.2 Proximal Policy Optimization (PPO)

PPO [9] has emerged as a significant algorithm in RL, designed to balance computational efficiency with the benefits of stable policy updates. PPO simplifies the complex optimization processes involved in earlier methods like TRPO, while maintaining robust performance levels.

3.7.2.1 PPO with KL Penalties

This approach modifies the objective function of PPO to include a penalty term based on the KL divergence. This penalty ensures controlled deviations from the old policy, aiding in maintaining stability. The objective function with the PPO Penalty is given by:

$$L(\theta) = \mathbb{E}\left[\frac{\pi_\theta(a|s)}{\pi_{\text{old}}(a|s)}A(s,a)\right] - \beta\mathbb{E}\left[D_{\text{KL}}\left(\pi_{\text{old}}(\cdot|s)||\pi_\theta(\cdot|s)\right)\right] \tag{3.12}$$

In this function, β is the coefficient determining the weight of the KL penalty. The balance between policy improvement and penalization of large deviations is crucial for effective policy updates. Intuitively, we see that there may not be any single value for β that works consistently across an entire training period, or between applications. Values for β are modified dynamically during training by comparing the divergence of the update to a target divergence, δ. If the observed divergence exceeds δ, we increase the penalty. If the observed divergence is too small, we decrease the penalty.

3.7.2.2 PPO with Clipped Objectives

An alternative variant of PPO introduces a clipping mechanism in the objective function to prevent large updates. The clipped objective is represented as:

$$L_{\text{clip}}(\theta) = \mathbb{E}\left[\min\left(r_t(\theta)A(s,a), \text{clip}(r_t(\theta), 1-\epsilon, 1+\epsilon)A(s,a)\right)\right] \tag{3.13}$$

Here, $r_t(\theta) = \frac{\pi_\theta(a|s)}{\pi_{\text{old}}(a|s)}$ is the ratio of new to old policy probabilities. The clipping function, $\text{clip}(r_t(\theta), 1-\epsilon, 1+\epsilon)$, limits this ratio, ensuring moderate policy updates, and maintaining stability in the learning process.

Both methods address the challenge of stabilizing policy updates in RL. While the PPO penalty method utilizes a KL divergence-based penalty, the PPO Clip method employs a clipping mechanism to achieve controlled policy updates. PPO's balance of performance, simplicity, and stability makes it a popular choice in various RL applications.

3.8 Conclusion

This chapter explored RL algorithms, from DQN to PPO, and revealed the significant evolutions in the field. Each algorithm introduced addresses specific challenges and limitations of its predecessors, leading to more efficient and stable learning processes.

DQN marked a significant advancement in RL by integrating deep learning with Q-learning. This advancement enabled the handling of high-dimensional state spaces. However, the efficiency was hindered by challenges like overestimation bias and limited applicability to discrete action spaces.

A2C demonstrated a dual-structure approach. It separated the policy execution (actor) and value estimation (critic). This separation allowed for more precise learning, particularly in environments containing complex or continuous action spaces. Limitations to A2C's sample inefficiency and inconsistency in policy updates necessitated further refinement.

TRPO was created to address these issues. TRPO introduced a "trust region" around the policy updates for stability. This methodology was adequate, but TRPO's computational complexity and overhead led to the development of PPO.

PPO stands out as superior due to its balance of computation efficiency and stable policy updates. It simplifies the optimization process introduced in TRPO, either through a KL penalty implementation or clipping mechanism, thus ensuring controlled policy deviation and increased performance.

Overall, the progression of advancements from DQN to PPO highlights the dynamic and fast-paced nature of RL research. Each algorithm and advancement builds upon the insightful foundations of the prior while attempting to overcome their shortcomings, contributing to the development of more sophisticated, efficient, and stable learning methodologies.

These evolutions enhance our understanding of complex decision-making processes and expand the applicability of RL in various real-world scenarios. The next chapter, "Motivation for Model Driven Penetration Testing," will explore some concrete examples of RL-based automation of the penetration testing methodologies introduced in Chapter 2, "Overview of Penetration Testing."

References

1 Cameron B Browne, Edward Powley, Daniel Whitehouse, Simon M Lucas, Peter I Cowling, Philipp Rohlfshagen, Stephen Tavener, Diego Perez, Spyridon Samothrakis, and Simon Colton. A survey of Monte Carlo tree search methods. *IEEE Transactions on Computational Intelligence and AI in games*, 4(1):1–43, 2012.

2 Longxin Lin. Self-improving reactive agents based on reinforcement learning, planning and teaching. *Machine Learning*, 8:293–321, 1992. URL https://api.semanticscholar.org/CorpusID:3248358.

3 Volodymyr Mnih, Koray Kavukcuoglu, David Silver, Alex Graves, Ioannis Antonoglou, Daan Wierstra, and Martin Riedmiller. Playing atari with deep reinforcement learning. 2013. URL http://arxiv.org/abs/1312.5602.

4 Athanasios S Polydoros and Lazaros Nalpantidis. Survey of model-based reinforcement learning: applications on robotics. *Journal of Intelligent & Robotic Systems*, 86(2):153–173, 2017.

5 Joe S Qin and Thomas A Badgwell. An overview of industrial model predictive control technology. In *AIche Symposium Series*, volume 93, pages 232–256. New York, NY: American Institute of Chemical Engineers, 1971–c2002. 1997.

6 Joshua Romoff. *Decomposing the Bellman equation in reinforcement learning*. McGill University (Canada), 2021.

7 John Schulman, Sergey Levine, Pieter Abbeel, Michael I Jordan, and Philipp Moritz. Trust region policy optimization. *ArXiv*, abs/1502.05477, 2015a. URL https://api.semanticscholar.org/CorpusID:16046818.

8 John Schulman, Philipp Moritz, Sergey Levine, Michael I Jordan, and Pieter Abbeel. High-dimensional continuous control using generalized advantage

estimation. *CoRR*, abs/1506.02438, 2015b. URL https://api.semanticscholar.org/CorpusID:3075448.

9 John Schulman, Filip Wolski, Prafulla Dhariwal, Alec Radford, and Oleg Klimov. Proximal policy optimization algorithms. *ArXiv*, abs/1707.06347, 2017a. URL https://api.semanticscholar.org/CorpusID:28695052.

10 John Schulman, Filip Wolski, Prafulla Dhariwal, Alec Radford, and Oleg Klimov. Proximal policy optimization algorithms. *arXiv preprint arXiv:1707.06347*, 2017b.

11 Miguel Suau, Elena Congeduti, Rolf A N Starre, Aleksander Czechowski, and Frans A Oliehoek. Influence-aware memory for deep reinforcement learning. *ArXiv*, abs/1911.07643, 2019. URL https://api.semanticscholar.org/CorpusID:208139426.

12 Martijn van Otterlo and Marco Wiering. *Reinforcement learning and Markov decision processes*, pages 3–42. Springer Berlin Heidelberg, Berlin, Heidelberg, 2012.

13 Junzi Zhang, Jongho Kim, Brendan O'Donoghue, and Stephen P Boyd. Sample efficient reinforcement learning with reinforce. *ArXiv*, abs/2010.11364, 2020. URL https://api.semanticscholar.org/CorpusID:225039878.

14 Dongbin Zhao, Haitao Wang, Kun Shao, and Yuanheng Zhu. Deep reinforcement learning with experience replay based on sarsa. In *2016 IEEE Symposium Series on Computational Intelligence (SSCI)*, pages 1–6, 2016. doi: 10.1109/SSCI.2016.7849837.

4

Motivation for Model-driven Penetration Testing

4.1 Introduction

Modern automated penetration testing uses rule-based procedures and model-checking concepts to search through all possible attacks on network models and identify those that violate some correctness or security property by generating an attack graph. By generating all possible attacks, modern top-down approaches inherently fail to isolate the relevant attacks that matter the most. Instead of finding these proverbial "needles in haystacks," they create more haystacks, or haystacks of needles: many variants of attack graphs grow exponentially with the number of hosts in a network. This weakness is exacerbated in other network architectures including 5G, Internet of Things (IoT), industrial control systems (ICS), and supervisory control, and data acquisition (SCADA) systems where networks may have thousands of hosts (or more) and evolve over time.

This has created a perception that the attack graph concept itself is inadequate, in turn hindering the automation of cyber testing. Recent reinforcement learning (RL) research (re)positions automated attack graph generation as a best practice in cyber defense by applying deep RL [17, 23, 24, 30, 43, 63]. RL is a subfield of artificial intelligence concerned with learning by interaction. Instead of generating exhaustive and succinct attack graphs that contain all and only attack paths that violate a given correctness or security property, RL agents learn optimal policies that link states to actions to optimize Markov decision processes (MDPs) modeled over the network models (or over attack graphs themselves), as depicted in Figure 4.1. These optimal policies can be used to find the individual attack paths that matter most, thereby aiding cyber analysts, operators, and automated-AI-enhanced workflows in timely decision support.

Reinforcement Learning for Cyber Operations: Applications of Artificial Intelligence for Penetration Testing, First Edition. Abdul Rahman, Christopher Redino, Dhruv Nandakumar, Tyler Cody, Sachin Shetty, and Dan Radke.

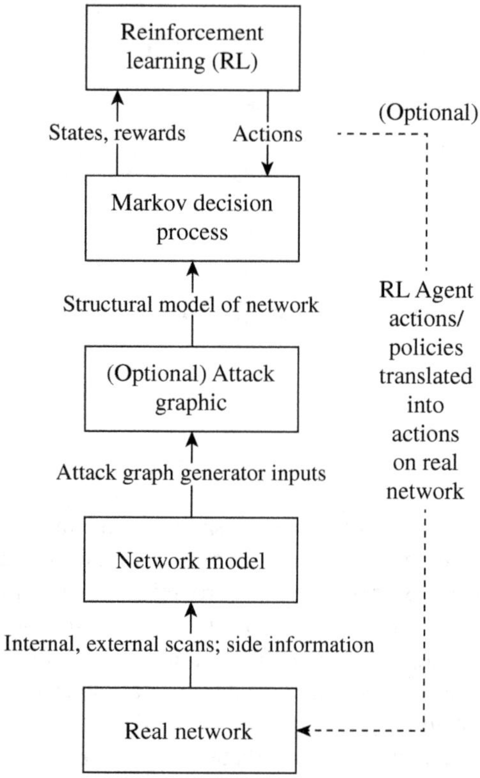

Figure 4.1 Layered processes with which RL agents interact.

In this chapter, we explore the technical limitations of existing methods that motivate the use of RL. Then, we discuss critiques of RL and MDP-based approaches to automating penetration testing. We then provide a survey of recent research into RL for penetration testing, identify related research trends, and propose *whole campaign emulation* (WCE) as a long-term, motivating goal for RL-based automation of penetration testing.

Recall from Chapter 3 that an RL agent interacts with an environment over a discrete number of time-steps by selecting an action at each time-step. In return, the environment returns the agent to a new state and a reward. Thus, the interaction between the agent and environment can be seen as a sequence of states and actions. When the agent reaches a terminal state, such as escalating privileges on a target host, the process stops. Formally speaking, the states, actions, admissible state-action pairs, state transition probabilities, and rewards comprise the MDP. Through interaction with MDP environments, RL agents learn to minimize the discounted sum of expected future rewards. Figure 4.1 visualizes RL-based penetration testing as a layered process.

4.2 Limits of Modern Attack Graphs

Barik et al. [6] trace the development of attack graphs first from privilege graphs [20], to state enumeration graphs [50], to scenario graphs [56], then from there jointly to exploit dependency graphs [2, 3], as used in topological vulnerability analysis (TVA), to logic programming graphs, as used in MulVal [48], and to multiple-prerequisite graphs, as used in NetSPA [32]. Privilege graphs were a precursor to attack graphs that modeled users, their privileges, and their vulnerabilities [20]. Deemed too user-centric, state enumeration graphs that modeled hosts, their configurations, and their vulnerabilities directly were introduced [50, 56]. Generating state enumeration graphs and closely related scenario graphs have an exponential growth rate with respect to the number of hosts in a network.

By using the monotonicity assumption [2], which specifies that privileges can only be gained, TVA approaches using exploit dependency graphs [3], and logic programming approaches like MulVal [48] are able to reduce the exponential growth rate to polynomial. But to do so, both model the vulnerabilities or the vulnerabilities on hosts directly, as opposed to the hosts themselves, and therefore do not embed or represent the entire network model. NetSPA is able to further reduce the growth rate to be log-linear and maintains the concept of host at the node-level in attack graphs [32]. But, NetSPA generates cyclical graphs, in contrast to the acyclical graphs generated by TVA, MulVal, and (exhaustive and succinct) model-checking approaches to state enumeration graphs. Table 4.1 summarizes the basic character of these varied approaches to attack graph generation.

There are a number of works in the literature that seek to extend the methods in Table 4.1. Topics include addressing completeness [47], complexity [27, 33], and what-if analyses [13, 65]. Current research into attack graph generation focuses on alert-driven methodology. This includes approaches that weigh completeness against alerts [28] and others that minimize the use of *a priori* expert knowledge [44]. Alert-driven attack graph generation can greatly reduce the scope of attack graph generation to relevant parts of networks and

Table 4.1 Attack graph generation methods.

Method	Growth rate in hosts n	Are nodes hosts?	Is the graph acyclic?
State-enumeration	$\mathcal{O}(2^n)$	Yes	No
Scenario	$\mathcal{O}(2^n)$	Partially	Yes
TVA	$\mathcal{O}(n^3)$	No	Yes
MulVal	$\mathcal{O}(n^2)$	No	Yes
NetSPA	$\mathcal{O}(n \log n)$	Sometimes	No

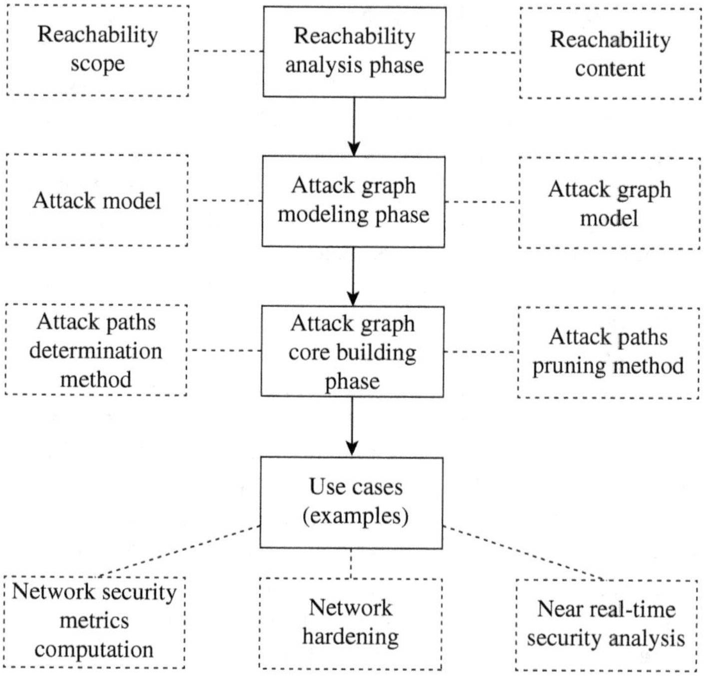

Figure 4.2 A taxonomy for attack graph generation and analysis.

is a promising area of research. Current efforts are tied more closely to attack graph visualization than to automated penetration testing with RL, however, for example, in the case of ransomware [43].

Generating and analyzing attack graphs is complex and merits its own taxonomy. Kaynar provides a general taxonomy for the basic workflow of using attack graphs [35], as shown in Figure 4.2. The basic workflow is that *reachability analysis* provides a network model, then an *attack graph model* must be chosen, e.g., from the aforementioned, and then the attack graph must be generated according to the chosen attack graph model and reachability analysis. This latter phase is termed the *core building phase*. The final phase is to perform analysis of the attack graph, i.e., to apply it to a use case. Kaynar breaks these phases into a number of subtypes, but even Kaynar admits that the taxonomy is unable to partition the literature, i.e., generation and analysis methods belong to multiple subtypes, as attack graph generation and analysis is a complex and technically detailed process.

4.2.1 Critiques of MDPs with Attack Graphs

The use of ML in cyber applications has broad, general challenges [4, 22, 59]. The use of RL with attack graphs has drawn specific criticism. Shmaryahu et al.

elaborate the difficulties assuming that MDPs modeled over attack graphs are fully observable, as well as with treating such difficulties with partially observable MDPs (POMDPs) [57]. In short, they prefer attack trees to attack graphs. Formally speaking, it is unclear that taking the more specific tree structure instead of a graph structure makes a significant difference regarding observability if the full attack graph or full attack tree is not provided to the RL agent as a given. Notions of observability will be discussed later in the chapter.

Other criticisms concern realism and using vulnerability metrics like the common vulnerability scoring system (CVSS) [34, 42]. Specifically, using the CVSS and metrics derived from other vulnerability databases helps automate and scale MDP construction. However, using vulnerability information exclusively provides a representation of networks that is highly biased toward known vulnerabilities [66]. Gangupantulu et al. proposed balancing this bias by introducing concepts of cyber terrain [19, 38] into MDPs [24], which demonstrates viable performance and is discussed later in the chapter.

4.2.2 Ontology-based Approaches

One alternative to dealing with the complexities and difficulties of attack graphs is to take an entirely different orientation to apply RL to computer networks. Ontology-based approaches to penetration testing [1, 31, 58, 61] can avoid using attack graphs using model-based RL. Agent-based ontologies can use a formal or semi-formal domain-specific language (DSL) to interact directly with software tools that in turn interact with networks [11, 12, 37]. In essence, this changes the network structure from an external input to an internal representation the RL agent must learn. Hybrid approaches can provide ontology-based RL agents with network structure as an input, e.g., via an attack graph, but naturally, this (re)introduces the challenges associated with constructing MDPs described later in this chapter.

4.3 RL for Penetration Testing

Recently, enabled by advances in deep learning, RL has seen broad application in cyber [45], and a revival of interest from the penetration testing community [10, 16, 23, 24, 26, 29, 46, 54, 60, 67–69]. Recently published literature on penetration testing with RL and attack graphs is summarized in Table 4.2.

Denis et al. define penetration testing as "a simulation of an attack to verify the security of a system or environment to be analyzed … through physical means utilizing hardware, or through social engineering" [21]. As presented in Chapter 2, if port scanning is looking through binoculars at a building to identify entry points, penetration testing is having someone actually break into the building

Table 4.2 Penetration Testing Using RL with Attack Graphs. From ref. [14], 2022, IEEE.

Authors	Network size	Model	Task
Yousefi et al.	44 vertices, 43 edges	MDP	Pathing
Schwartz and Kurniawati	50 hosts	MDP	Pathing
Ghanem and Chen	100 hosts	POMDP	Unspecified
Chowdary et al.	109 vertices, edges unknown	MDP	Pathing
Hu et al.	44 vertices, 52 edges	MDP	Pathing
Gangupantulu et al.	955 vertices, 2350 edges	MDP	Pathing
Gangupantulu et al.	1617 vertices, 4331 edges	MDP	Crown jewel analysis
Nguyen et al.	1024 hosts	MDP	Pathing
Zhou et al.	17 hosts	MDP	Pathings
Zennaro and Erodi	Not reported	MDP	Capture the flag
Tran et al.	900 hosts	MDP	Pathing
Cody et al.	26 hosts	MDP	Exfiltration

Source: From Cody [14], 2022, IEEE.

to determine the viability of these entry points. Penetration testing and static analysis combine to make up vulnerability detection and analysis [5, 8, 55].

A unique AI-focused contrast to the discussion in Chapter 2 on penetration testing historically posits that pentesting models have taken the form of the flaw hypothesis model [49, 64], attack trees [51, 52], and attack graphs [39]. Each approach has its own strengths. But attack graphs are particularly well-suited for RL. In fact, the earliest use of RL with attack graphs may be the use of the value iteration algorithm by Sheyner et al. to do probabilistic reliability analysis in their seminal work on scenario graphs [56].

Recall from Chapter 3, that an agent is "typically" considered to interact with an environment $\mathcal{E}$ over a discrete number of time-steps by selecting an action a_t at time-step t from the set of actions A. In return, the environment $\mathcal{E}$ returns to the agent a new state s_{t+1} and reward r_{t+1}. Thus, the interaction between the agent and environment $\mathcal{E}$ can be seen as a sequence $s_1, a_1, s_2, a_2, \dots, a_{t-1}, s_t$. When the agent reaches a terminal state, the process stops. When $\mathcal{E}$ is a finite MDP, it is a tuple $\langle S, A, \Phi, P, R \rangle$, where S is a set of states, A is a set of actions, $\Phi \subset S \times A$ is the set of admissible state-action pairs, $P : \Phi \times S \rightarrow [0, 1]$ is the transition probability function, and $R : \Phi \rightarrow \mathbb{R}$ is the expected reward function where $\mathbb{R}$ is the set of real numbers. $P(s, a, s')$ denotes the transition probability from state s to state s' under action a, and $R(s, a)$ denotes the expected reward from taking action a in state s. The goal of learning is to find a policy $\pi \in \Pi$, where $\Pi : S \rightarrow A$, which maximizes the sum of discounted future rewards.

As noted in Table 4.2, other than Ghanem and Chen [26], the authors in the RL with attack graphs literature use fully observable MDPs to model networks. Many authors use the CVSS to furnish their MDPs. Yousefi et al. provide the earliest work doing so in deep RL for penetration testing [67]. Hu et al. extend the use of the CVSS by proposing to use exploitability scores weight rewards [29]. Gangupantulu et al. [23, 24] and Cody et al. [16] explicitly extend the methods of Hu et al. with concepts of cyber terrain as articulated by Conti and Raymond [19]. Gangupantulu et al. advocate defining models of terrain in terms of the rewards and transition probabilities of MDPs, first in the case of firewalls as obstacles [24], then in the case of lateral pivots nearby key terrain [23]. Cody et al. apply these concepts to exfiltration [16]. Other authors either handcraft the MDP or do not remark on how its components are estimated.

Many authors apply generic deep Q learning (DQN) [40, 41] to solve point-to-point network traversal tasks, termed "pathing" in Table 4.2 [10, 24, 29, 54, 67]. Typically the terminal state is unknown and solutions take the form of individual paths. Others develop domain-specific modifications for deep RL including the double agent architecture (DAA) [46], a hierarchical action decomposition approach [60], and various improvements to DQN termed NDSPI-DQN [69]. Another line of research focuses on developing more specific penetration testing tasks than "pathing." A number of authors define more specific tasks by reward engineering and other modifications to the MDP including formulations of capture the flag [68], crown jewel analysis [23], and discovering exfiltration paths [16].

4.4 Modeling MDPs

For RL-based penetration testing, the states are (often) the hosts and their conditions (e.g., reachable, discovered, and access level) and the actions are (often) a combination of scans, exploits, and privilege escalations. There is less consistency in how rewards are represented [14]. Rewards often consist of (1) costs or penalties (e.g., penalty per step or cost to use an exploit), (2) lump-sum terminal rewards (e.g., for escalating privileges on a target host), and (3) intermediate rewards associated with cyber terrain that draws from vulnerability databases, knowledge of network configuration and defenses, and expert knowledge. Lastly, the state transition probabilities are (often) associated with the difficulty of executing an action (e.g., scan, exploit, or escalation), which can be drawn from "exploitability" scores offered by vulnerability databases.

The third category of rewards – intermediate rewards corresponding to cyber terrain – is worth underscoring. Ultimately, automated penetration testing should find attack paths that matter. A key aspect of finding such "paths that matter" is ensuring identified attack paths match both the incentives as well as the techniques, tactics, and procedures (TTPs) of adversaries. While TTPs historically have been modeled using the action space (i.e., available vulnerabilities, exploits), in modern attack graph generation approaches, incorporating models of cyber terrain into MDPs over network models (and attack graphs) offers an alternative approach.

Using concepts of cyber terrain corresponding to intelligence preparation of the battlefield, rewards and transition probabilities can be crafted, e.g., to represent firewalls as obstacles [24] and to associate intrusion detection systems with field of fire [30]. By incorporating cyber terrain, the task of emulating realistic adversaries becomes shared among several different MDP engineering efforts. These can be treated as a series of layered, modeling processes which take a generic MDP, imbue it with cyber terrain, tailor it to a given adversary, and then modify it to fit various tasks.

This concept of layered MDPs for penetration testing was outlined as a standard reference architecture after a survey of the literature [14]. While the broader field has focused on generic, point-to-point network traversal and privilege escalation using RL, recent developments propose more tailored tasks like crown jewel analysis [23], discovering exfiltration paths [16], and surveillance detection routes (SDR) [30]. To create these tailored tasks, authors vary different aspects of the MDP in the various ways described above.

MDP engineering and its role in creating tailored tasks is one of the three major, current, and academic research thrusts. The other two major thrusts consider partial observability and scalable action spaces. It can be straightforwardly argued that POMDPs provide more natural models for penetration testing than fully observable MDPs because a complete view of the tested network is almost never available in practice [57]. POMDPs relax the observability assumption of MDPs, and, thus, prior work on MDPs can be applied to POMDPs as well.

The scalable action space thrust is another practical consideration. Action spaces grow quickly since each additional host adds many possible actions. The actions are typically hierarchical, e.g., there are scans; scans can be divided into subnetwork, operating system, service, and process scans; each of those subclasses may have several tools for execution, and each of those tools may have several parameterizations. The hierarchical nature of penetration testing actions has been exploited by a variety of multi-agent approaches [60]. Typically, different agents are tasked with learning different classes (e.g., scanning vs. exploiting) or with learning different levels (e.g., which scanning subclass to select vs. which tool parameterization to select), and are jointly optimized using a scheme for combining the agents' rewards.

Across the partial observability, scalability, and MDP engineering research thrusts, there is not a coherent research objective. The concept of WCE offers a generic research objective which can align these various research thrusts. MDP engineering, partial observability, and scalable action spaces, woven together by the unifying concept of WCE, can undergird the next generation of automated penetration testing systems.

4.4.1 Whole Campaign Emulation

Recently, WCE was defined as a challenging problem and framework for automated penetration testing that treats RL as a means of emulating all stages of attack campaigns. Cody, Beling, and Freeman define WCE formally as the optimization of a set of MDPs corresponding to different stages of attack campaigns [15, 18]. Put simply, WCE shifts the focus of automated penetration testing from entirely holistic approaches (entire attack campaign described by a single MDP) to a combination of tailored, differentiated approaches (stages of attack campaign described by different MDPs), as shown in Figures 4.3 and 4.4.

The practical motivation is straightforward – the various stages have different requirements but occur in the same network. Consider, exfiltration tasks are sensitive to the rate of packet flows in a way that reconnaissance tasks may not be, suggesting that one MDP may include time in the state while another may not. Yet further, command and control tasks may require time to be included in the state as well as notions of communication signals. While to an academic these are transformations on a common MDP, in practice, the resulting differences in the information needed to construct these MDPs may lead to stage-specific RL-based cyber tools.

Such modularization allows for methods for easier-to-solve stages to advance through levels of technology readiness faster than methods for harder-to-solve stages. This follows a classic adage of AI [62]: "do not solve a more general problem in order to solve a specific problem." It also follows the traditional path of process automation in cyber workflows. Overtime, new automated tools act as force multipliers for cyber operators, not replacements for cyber operators. At the same time, by continuing to pursue WCE, fundamental research remains aligned toward automating the entire campaign.

This alignment can be seen in a different but motivating research area termed whole brain emulation (WBE). WBE posits that artificial general intelligence can

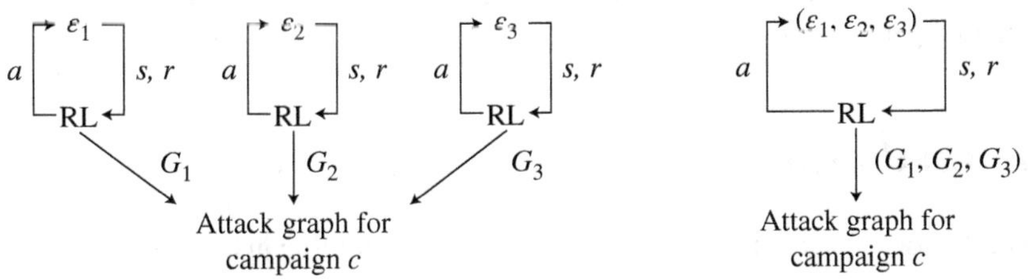

Figure 4.3 A depiction of whole campaign emulation (WCE).

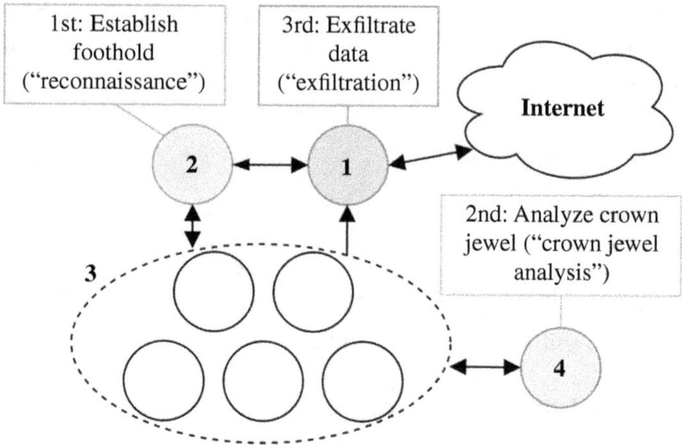

Figure 4.4 An example of whole campaign emulation (WCE). Source: Cody et al. [18] © [2023] IEEE.

be achieved by digitally emulating a human brain [36]. This lofty vision has led to grounded understanding of the current state of the art. The lack of capability of sensors, algorithms, and brain models, e.g., in terms of latency, information processing capacity, abstraction and fidelity, respectively, is exposed by the unifying research objective of WBE. WCE offers an analogous benchmark for RL-based automated penetration testing. The progress of various research thrusts like MDP engineering, partial observability, and scalability can be judged by the lofty vision of automated penetration testing offered by WCE.

It may be the case that, in the long run, WCE will be solved by removing the barriers between stages and working backward toward a single-unified MDP with a sparse terminal reward. If so, to some, emphasizing the role of specific stages in WCE may seem like an unnecessary step. However, without a rigorous concept for WCE, the research community lacks a framework for judging if WCE has been achieved in the first place. It may also be the case that, in the long-run, stage-specific RL-based automation solutions continue to diversify into their own niches. In such a case, with WCE as a unifying research objective, it is clear how progress in their segregated development contributes to the overall objective of automated penetration testing.

4.5 Conclusion

The motivation for this manuscript can be seen by sampling just a few of our key contributions within [17, 23, 24, 30, 63]:

- In [24], we contribute methodology for building OAKOC cyber terrain into MDP models of attack graphs. In [24], we apply our methodology to RL for penetration testing by treating firewalls as cyber terrain obstacles in an example network that is at least an order of magnitude larger than the networks used by previous authors [7, 9, 25, 26, 29, 54, 67].
- In [24], we extend the literature on using CVSS scores to construct attack graphs and MDPs as well as the literature on RL for penetration testing. In [23], we present a novel method for using reinforcement learning in crown jewel analysis, termed CJA-RL, that incorporates network structure, cyber terrain, and path structure to enhance operator workflow by automating the construction and analysis of attack campaigns from various entry points or initial nodes to a CJ's 2-hop network.
- In [17], we provide an approach for modeling service-based defensive cyber terrain in dynamic models of attack graphs was proposed, including an RL-based algorithm for discovering the top-N exfiltration paths in an attack graph.
- In [30], we propose an MDP formulation and a new algorithm, SDR-RL, that uses a warm-up phase and penalty scaling to control the asymmetry between the number of host services scanned and the number of defensive terrain encountered. This emulates the asymmetry sought by human operators when conducting network reconnaissance generally and SDR in particular. Also, in [30], we extend the DAA of Nguyen et al. [46], which originally used standard DQN, with actor-critic (A2C) and proximal-policy optimization (PPO) algorithms [53].

This manuscript provides methods across the campaign that all leverage a common interest in cyber terrain and work not by identifying all possible attack paths, but rather by identifying the few attack paths that matter. The following sections discuss how RL impacts the detection of adverse behavior for various cyber security areas that include cyber terrain, crown jewels discovery, analysis of exfiltration paths, and advanced reconnaissance pathway discovery (i.e., SDR).

References

1. F Abdoli, N Meibody, and R Bazoubandi. An attacks ontology for computer and networks attack. In *Innovations and advances in computer sciences and engineering*, pages 473–476. Springer, 2010.
2. Paul Ammann, Duminda Wijesekera, and Saket Kaushik. Scalable, graph-based network vulnerability analysis. In *Proceedings of the 9th ACM Conference on Computer and Communications Security*, pages 217–224, 2002.
3. Paul Ammann, Joseph Pamula, Ronald Ritchey, and Julie Street. A host-based approach to network attack chaining analysis. In *21st Annual Computer Security Applications Conference (ACSAC'05)*, pages 10pp. IEEE, 2005.

4 Giovanni Apruzzese, Michele Colajanni, Luca Ferretti, Alessandro Guido, and Mirco Marchetti. On the effectiveness of machine and deep learning for cyber security. In *2018 10th International Conference on Cyber Conflict (CyCon)*, pages 371–390. IEEE, 2018.

5 Aileen G Bacudio, Xiaohong Yuan, Bei-Tseng B Chu, and Monique Jones. An overview of penetration testing. *International Journal of Network Security and Its Applications*, 3(6):19, 2011.

6 Mridul S Barik, Anirban Sengupta, and Chandan Mazumdar. Attack graph generation and analysis techniques. *Defence Science Journal*, 66(6): 559, 2016.

7 Sujita Chaudhary, Austin O'Brien, and Shengjie Xu. Automated post-breach penetration testing through reinforcement learning. In *2020 IEEE Conference on Communications and Network Security (CNS)*, pages 1–2. IEEE, 2020.

8 Brian Chess and Gary McGraw. Static analysis for security. *IEEE Security and Privacy*, 2(6):76–79, 2004.

9 Ankur Chowdary, Dijiang Huang, Jayasurya S Mahendran, Daniel Romo, Yuli Deng, and Abdulhakim Sabur. Autonomous security analysis and penetration testing. 2020.

10 Ankur Chowdhary, Dijiang Huang, Jayasurya S Mahendran, Daniel Romo, Yuli Deng, and Abdulhakim Sabur. Autonomous security analysis and penetration testing. In *2020 16th International Conference on Mobility, Sensing and Networking (MSN)*, pages 508–515. IEEE, 2020.

11 Ge Chu and Alexei Lisitsa. Penetration testing for internet of things and its automation. In *2018 IEEE 20th International Conference on High Performance Computing and Communications; IEEE 16th International Conference on Smart City; IEEE 4th International Conference on Data Science and Systems (HPCC/SmartCity/DSS)*, pages 1479–1484. IEEE, 2018.

12 Ge Chu and Alexei Lisitsa. Ontology-based automation of penetration testing. In *ICISSP*, pages 713–720, 2020.

13 Matthew Chu, Kyle Ingols, Richard Lippmann, Seth Webster, and Stephen Boyer. Visualizing attack graphs, reachability, and trust relationships with navigator. In *Proceedings of the Seventh International Symposium on Visualization for Cyber Security*, pages 22–33, 2010.

14 Tyler Cody. A layered reference model for penetration testing with reinforcement learning and attack graphs. In *2022 IEEE 29th Annual Software Technology Conference (STC)*, pages 41–50. IEEE, 2022.

15 Tyler Cody, Peter Beling, and Laura Freeman. Towards continuous cyber testing with reinforcement learning for whole campaign emulation. In *2022 IEEE AUTOTESTCON*, pages 1–5. IEEE, 2022.

16 Tyler Cody, Abdul Rahman, Christopher Redino, Lanxiao Huang, Ryan Clark, Akshay Kakkar, Deepak Kushwaha, Paul Park, Peter Beling, and Edward Bowen. Discovering exfiltration paths using reinforcement learning with attack graphs. *arXiv preprint arXiv:2201.12416*, 2022.

17 Tyler Cody, Abdul Rahman, Christopher Redino, Lanxiao Huang, Ryan Clark, Akshay Kakkar, Deepak Kushwaha, Paul Park, Peter Beling, and Edward Bowen. Discovering exfiltration paths using reinforcement learning with attack graphs. In *2022 IEEE Conference on Dependable and Secure Computing (DSC)*, pages 1–8, 2022.

18 Tyler Cody, Emma Meno, Peter Beling, and Laura Freeman. Whole campaign emulation with reinforcement learning for cyber test. *IEEE Instrumentation and Measurement Magazine*, 26(5):25–30, 2023.

19 Greg Conti and David Raymond. *On cyber: towards an operational art for cyber conflict*. Kopidion Press, 2018.

20 Marc Dacier and Yves Deswarte. Privilege graph: an extension to the typed access matrix model. In *European Symposium on Research in Computer Security*, pages 319–334. Springer, 1994.

21 Matthew Denis, Carlos Zena, and Thaier Hayajneh. Penetration testing: Concepts, attack methods, and defense strategies. In *2016 IEEE Long Island Systems, Applications and Technology Conference (LISAT)*, pages 1–6. IEEE, 2016.

22 James B Fraley and James Cannady. The promise of machine learning in cybersecurity. In *SoutheastCon 2017*, pages 1–6. IEEE, 2017.

23 Rohit Gangupantulu, Tyler Cody, Abdul Rahman, Christopher Redino, Ryan Clark, and Paul Park. Crown jewels analysis using reinforcement learning with attack graphs. *2021 IEEE Symposium Series on Computational Intelligence (SSCI)*, pages 1–6, 2021.

24 Rohit Gangupantulu, Tyler Cody, Paul Park, Abdul Rahman, Logan Eisenbeiser, Dan Radke, and Ryan Clark. Using cyber terrain in reinforcement learning for penetration testing. In *2022 IEEE International Conference on Omni-layer Intelligent Systems (COINS)*, pages 1–8, 2022.

25 Mohamed C Ghanem and Thomas M Chen. Reinforcement learning for intelligent penetration testing. In *2018 Second World Conference on Smart Trends in Systems, Security and Sustainability (WorldS4)*, pages 185–192. IEEE, 2018.

26 Mohamed C Ghanem and Thomas M Chen. Reinforcement learning for efficient network penetration testing. *Information*, 11(1):6, 2020.

27 John Homer, Ashok Varikuti, Xinming Ou, and Miles A McQueen. Improving attack graph visualization through data reduction and attack grouping. In *International Workshop on Visualization for Computer Security*, pages 68–79. Springer, 2008.

28 Hao Hu, Jing Liu, Yuchen Zhang, Yuling Liu, Xiaoyu Xu, and Jinglei Tan. Attack scenario reconstruction approach using attack graph and alert data mining. *Journal of Information Security and Applications*, 54:102522, 2020.

29 Zhenguo Hu, Razvan Beuran, and Yasuo Tan. Automated penetration testing using deep reinforcement learning. In *2020 IEEE European Symposium on Security and Privacy Workshops (EuroS&PW)*, pages 2–10. IEEE, 2020.

30 Lanxiao Huang, Tyler Cody, Abdul Rahman, Christopher Redino, Ryan Clark, Akshay Kakkar, Deepak Kushwaha, Paul Park, Peter Beling, and Edward Bowen. Exposing surveillance detection routes via reinforcement learning, attack graphs, and cyber terrain. In *2022 21st IEEE International Conference on Machine Learning and Applications (ICMLA)*, pages 1–8, 2022.

31 Michael Iannacone, Shawn Bohn, Grant Nakamura, John Gerth, Kelly Huffer, Robert Bridges, Erik Ferragut, and John Goodall. Developing an ontology for cyber security knowledge graphs. In *Proceedings of the 10th Annual Cyber and Information Security Research Conference*, pages 1–4, 2015.

32 Kyle Ingols, Richard Lippmann, and Keith Piwowarski. Practical attack graph generation for network defense. In *2006 22nd Annual Computer Security Applications Conference (ACSAC'06)*, pages 121–130. IEEE, 2006.

33 Kyle Ingols, Matthew Chu, Richard Lippmann, Seth Webster, and Stephen Boyer. Modeling modern network attacks and countermeasures using attack graphs. In *2009 Annual Computer Security Applications Conference*, pages 117–126. IEEE, 2009.

34 Pontus Johnson, Robert Lagerström, Mathias Ekstedt, and Ulrik Franke. Can the common vulnerability scoring system be trusted? A Bayesian analysis. *IEEE Transactions on Dependable and Secure Computing*, 15(6):1002–1015, 2016.

35 Kerem Kaynar. A taxonomy for attack graph generation and usage in network security. *Journal of Information Security and Applications*, 29:27–56, 2016.

36 Randal Koene and Diana Deca. Whole brain emulation seeks to implement a mind and its general intelligence through system identification. *Journal of Artificial General Intelligence*, 4(3):1–9, 2013.

37 Ryusei Maeda and Mamoru Mimura. Automating post-exploitation with deep reinforcement learning. *Computers and Security*, 100:102108, 2021.

38 Eviatar Matania and Eldad Tal-Shir. Continuous terrain remodelling: gaining the upper hand in cyber defence. *Journal of Cyber Policy*, 5(2):285–301, 2020.

39 James P McDermott. Attack net penetration testing. In *Proceedings of the 2000 Workshop on New Security Paradigms*, pages 15–21, 2001.

40 Volodymyr Mnih, Koray Kavukcuoglu, David Silver, Alex Graves, Ioannis Antonoglou, Daan Wierstra, and Martin Riedmiller. Playing atari with deep reinforcement learning. *arXiv preprint arXiv:1312.5602*, 2013.

41 Volodymyr Mnih, Koray Kavukcuoglu, David Silver, Andrei A Rusu, Joel Veness, Marc G Bellemare, Alex Graves, Martin Riedmiller, Andreas K Fidjeland, Georg Ostrovski, et al. Human-level control through deep reinforcement learning. *Nature*, 518(7540):529–533, 2015.

42 Nuthan Munaiah and Andrew Meneely. Vulnerability severity scoring and bounties: why the disconnect? In *Proceedings of the 2nd International Workshop on Software Analytics*, pages 8–14, 2016.

43 Sathvik Murli, Dhruv Nandakumar, Prabhat Kushwaha, Cheng Wang, Christopher Redino, Abdul Rahman, Shalini Israni, Tarun Singh, and Edward Bowen. Cross-temporal detection of novel ransomware campaigns: a multi-modal alert approach. *arXiv:2309.00700 [cs.CR]*, 2023.

44 Azqa Nadeem, Sicco Verwer, Stephen Moskal, and Shanchieh J Yang. Alert-driven attack graph generation using s-PDFA. *IEEE Transactions on Dependable and Secure Computing*, 19(2):731–46, 2021.

45 Thanh T Nguyen and Vijay J Reddi. Deep reinforcement learning for cyber security. *arXiv preprint arXiv:1906.05799*, 2019.

46 Hoang V Nguyen, Songpon Teerakanok, Atsuo Inomata, and Tetsutaro Uehara. The proposal of double agent architecture using actor-critic algorithm for penetration testing. In *ICISSP*, pages 440–449, 2021.

47 Peng Ning, Dingbang Xu, Christopher G Healey, and Robert St Amant. Building attack scenarios through integration of complementary alert correlation method. In *NDSS*, volume 4, pages 97–111, 2004.

48 Xinming Ou, Wayne F Boyer, and Miles A McQueen. A scalable approach to attack graph generation. In *Proceedings of the 13th ACM Conference on Computer and Communications Security*, pages 336–345, 2006.

49 Charles P Pfleeger, Shari L Pfleeger, and Mary F Theofanos. A methodology for penetration testing. *Computers and Security*, 8(7):613–620, 1989.

50 Cynthia Phillips and Laura P Swiler. A graph based system for network-vulnerability analysis. In *Proceedings of the 1998 Workshop on New Security Paradigms*, pages 71–79, 1998.

51 Chris Salter, O S Saydjari, Bruce Schneier, and Jim Wallner. Toward a secure system engineering methodology. In *Proceedings of the 1998 Workshop on New Security Paradigms*, pages 2–10, 1998.

52 Bruce Schneier. Attack trees. *Dr. Dobb's Journal*, 24(12):21–29, 1999.

53 John Schulman, Filip Wolski, Prafulla Dhariwal, Alec Radford, and Oleg Klimov. Proximal policy optimization algorithms. *arXiv preprint arXiv:1707.06347*, 2017.

54 Jonathon Schwartz and Hanna Kurniawati. Autonomous penetration testing using reinforcement learning. *arXiv preprint arXiv:1905.05965*, 2019.

55 Sugandh Shah and Babu M Mehtre. An overview of vulnerability assessment and penetration testing techniques. *Journal of Computer Virology and Hacking Techniques*, 11(1):27–49, 2015.

56 Oleg Sheyner, Joshua Haines, Somesh Jha, Richard Lippmann, and Jeannette M Wing. Automated generation and analysis of attack graphs. In *Proceedings 2002 IEEE Symposium on Security and Privacy*, pages 273–284. IEEE, 2002.

57 Dorin Shmaryahu, G Shani, J Hoffmann, and M Steinmetz. Constructing plan trees for simulated penetration testing. In *The 26th International Conference on Automated Planning and Scheduling*, volume 121, 2016.

58 Andrew Simmonds, Peter Sandilands, and Louis Van Ekert. An ontology for network security attacks. In *Asian Applied Computing Conference*, pages 317–323. Springer, 2004.

59 Sumit Soni and Bharat Bhushan. Use of machine learning algorithms for designing efficient cyber security solutions. In *2019 2nd International Conference on Intelligent Computing, Instrumentation and Control Technologies (ICICICT)*, volume 1, pages 1496–1501. IEEE, 2019.

60 Khuong Tran, Ashlesha Akella, Maxwell Standen, Junae Kim, David Bowman, Toby Richer, and Chin-Teng Lin. Deep hierarchical reinforcement agents for automated penetration testing. *arXiv preprint arXiv:2109.06449*, 2021.

61 John Pinkston, Anupam Joshi, Tim Finin. A target-centric ontology for intrusion detection. In *Workshop on Ontologies in Distributed Systems, Held at the 18th International Joint Conference on Artificial Intelligence*, 2003. https://ebiquity.umbc.edu/paper/html/id/63/

62 Vladimir Vapnik. *The nature of statistical learning theory*. Springer Science & Business Media, 1999.

63 Cheng Wang, Akshay Kakkar, Chris Redino, Abdul Rahman, S Ajinsyam, Ryan Clark, Daniel Radke, Tyler Cody, Lanxiao Huang, and Edward Bowen. Discovering command and control (C2) channels using reinforcement learning (RL). *Submitted IEEE SouthEastCon 2023*, 2023.

64 Clark Weissman. Penetration testing. *Information Security: An Integrated Collection of Essays*, 6:269–296, 1995.

65 Leevar Williams, Richard Lippmann, and Kyle Ingols. Garnet: a graphical attack graph and reachability network evaluation tool. In *International Workshop on Visualization for Computer Security*, pages 44–59. Springer, 2008.

66 Jinyu Wu, Lihua Yin, and Yunchuan Guo. Cyber attacks prediction model based on bayesian network. In *2012 IEEE 18th International Conference on Parallel and Distributed Systems*, pages 730–731. IEEE, 2012.

67 Mehdi Yousefi, Nhamo Mtetwa, Yan Zhang, and Huaglory Tianfield. A reinforcement learning approach for attack graph analysis. In *2018 17th IEEE International Conference On Trust, Security and Privacy In Computing and Communications/12th IEEE International Conference On Big Data Science and Engineering (TrustCom/BigDataSE)*, pages 212–217. IEEE, 2018.

68 Fabio M Zennaro and Laszlo Erdodi. Modeling penetration testing with reinforcement learning using capture-the-flag challenges: trade-offs between model-free learning and a priori knowledge. *arXiv preprint arXiv:2005.12632*, 2020.

69 Shicheng Zhou, Jingju Liu, Dongdong Hou, Xiaofeng Zhong, and Yue Zhang. Autonomous penetration testing based on improved deep Q-network. *Applied Sciences*, 11(19):8823, 2021.

5

Operationalizing RL for Cyber Operations

5.1 A High-Level Architecture

The challenge of real-world applications of artificial intelligence (AI) and machine learning (ML) is that they need representations of reality that are well-grounded to the system within which they operate. This is known as the *grounding* problem [11, 16]. The application of reinforcement learning (RL) to automate penetration testing is no different. Attack graphs are commonly used representations in the academic literature [1, 14, 21]. But how those attack graphs are actively grounded to real networks has been understudied in the literature from the perspective of RL. Leaving the grounding problem unaddressed puts the reliability of proposed RL-based cybersecurity solutions into serious question. This paper presents a layered reference model for how RL agents are grounded to real networks, as shown in Figure 5.1.

How exhaustive or succinct should attack graphs that RL agents interact with be? One concept of an attack graph is that it consists of all and only those paths from an initial state that reach terminal or success states, e.g., by violating a security or correctness property of a network model [17]. Exhaustive and succinct attack graphs like these, when used with RL, keep their exponential and polynomial growth rates (with respect to the number of hosts in a network) in order to turn the RL task into a needle-in-the-haystack problem of finding realistic attack paths or subgraphs within the graph.

This bottlenecks the run-time of RL and directly violates a classic adage from learning theory, "do not solve a specific problem by first solving a more general problem" [19]. RL agents for penetration testing most often need to find a small set of realistic attack paths or subgraphs matching an attack template, playbook, adversary profile, etc.; to do so, they do not need to find every possible attack path first.

Reinforcement Learning for Cyber Operations: Applications of Artificial Intelligence for Penetration Testing, First Edition. Abdul Rahman, Christopher Redino, Dhruv Nandakumar, Tyler Cody, Sachin Shetty, and Dan Radke.

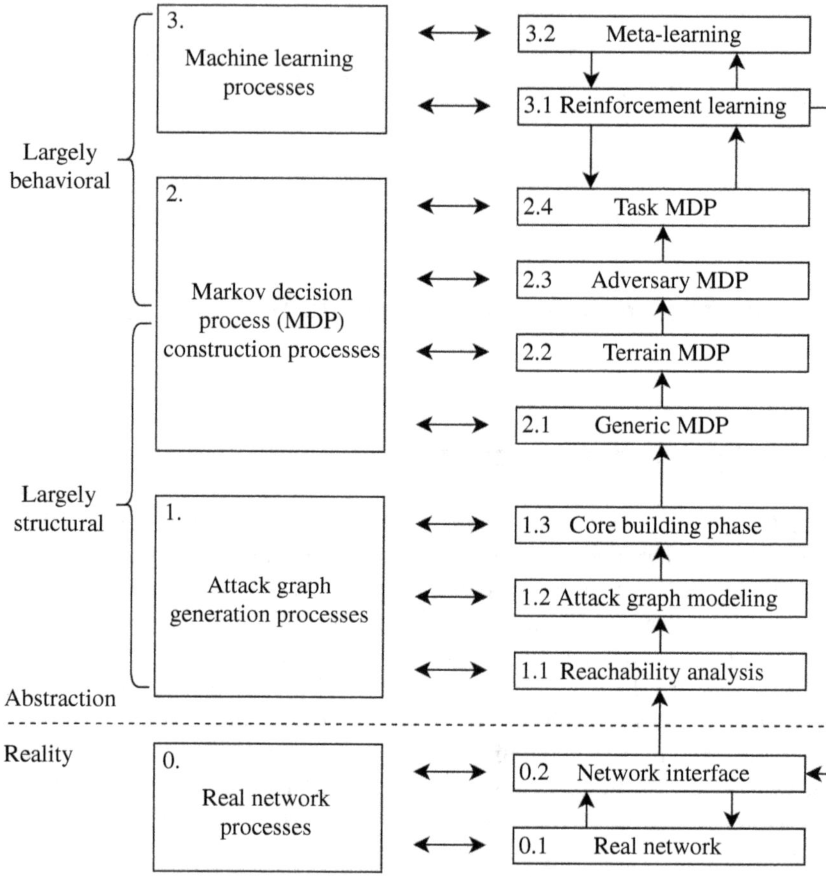

Figure 5.1 The layered reference model for RL with attack graphs (LRM-RAG). Structure and behavior refer to network (and path) structure and behavior. Source: From Cody [6], 2022, IEEE.

At the very least, formulating the RL agent's learning task as finding a subgraph within an exhaustive and succinct attack graph is an indirect formulation – the agent is more directly tasked with finding one or more realistic subgraphs within a network satisfying a goal (e.g., violating a security or correctness property of the tested network). Such an approach assumes a full attack graph has already been generated and given to the agent, and, therefore, presupposes that generating such a graph is a necessary part of solution methods.

It is not clear how the techniques developed and lessons learned from classical attack graph generation techniques like model checking [13] and logic programming [15] apply when RL agents, as opposed to human operators, will be interacting with the attack graphs. For example, should RL agents handle attack graph generation themselves? Or be confined to analysis? Or should they do both in an end-to-end manner? Answering these questions requires a broad view of the defining challenges of using RL with attack graphs.

Layered reference model for RL with attack graphs (LRM-RAG) provides a framing for these defining challenges in the form of a layered reference model [6]. The model is developed from a systems perspective rooted in the AI and ML character of using RL with attack graphs. LRM-RAG is defined and used to consider key challenges. Challenges are organized into three questions:

- How can RL interact with real networks in real time?
- How can RL emulate realistic adversary behavior?
- How can RL deal with unstable/evolving networks?

LRM-RAG can be summarized as follows:

Information from the real network is abstracted into an attack graph by the attack graph generation processes. This attack graph is a largely structural representation of vulnerabilities, their pre- and post-conditions, and (sometimes) hosts. To formulate an environment for the RL agent, a Markov decision process (MDP) is modeled over the generated attack graph using the MDP construction processes. The result of this is a layering of behavior and additional structure on top of the attack graph. The RL agent learns by interacting with the MDP, and as tasks, adversaries, and terrain change, the RL agent can transfer learn (i.e., share knowledge) or meta-learn (i.e., learn-to-learn) and thereby reapply lessons learned across penetration testing activities. The RL agent's interaction with the MDP is interpreted by the network interface to realize penetration testing and its outcomes in reality.

5.2 Layered Reference Model

LRM-RAG is constructed from general to specific, as shown when reading left-to-right in Figure 5.1. At the most general, a distinction is made between reality and the abstractions used to model it. Reality corresponds to the real network where penetration testing is applied. Attack graphs, MDPs, and RL are abstractions that model those networks, and they model structural and behavioral phenomena.

Structure is modeled by generating an attack graph and constructing an MDP. Next, behavior is modeled by enriching the transition probabilities and reward in the MDP and by training RL agents to interact with it. Then, informed by interacting with MDPs, RL agents can interact with real networks or inform interactions with real networks. RL agents observe outcomes of their actions in real networks or those they inform as filtered by the processes producing MDPs.

Table 5.1 Table of classification of elements in LRM-RAG.

Reality	**Abstraction**		
Layer 1 *Real network processes*	Layer 2 *Attack graph generation processes*	Layer 3 *MDP construction processes*	Layer 4 *Machine learning processes*
1.1 Real network	2.1 Reachability analysis	3.1 Generic MDP	4.1 Reinforcement learning
1.2 Network interface	2.2 Attack graph modeling	3.2 Terrain MDP	4.2 Meta-learning
	2.3 Core building phase	3.3 Adversary MDP	
		3.4 Task MDP	

LRM-RAG gives a heterarchical view of automating penetration testing with RL and attack graphs. It contrasts the hierarchical (feed-forward) view offered by Kaynar [14] depicted in Figure 4.2. The elements of LRM-RAG are classified in Table 5.1, and in the following, each element is defined explicitly.

5.2.1 Real Network Processes

The base layer in LRM-RAG is termed *real network processes* and it consists of the real network and its interface.

Definition 5.1 Real Network.
The *real network* is the (computer) network where penetration testing is applied.

The real network may be uncertain, dynamic, and only partially observable. Defining the real network is more of a general scoping and place-setting activity than an exact specification of network properties.

Definition 5.2 Network Interface.
The *network interface* is the means for interaction with the real network.

It has a two-way connection with the real network. It can be entirely automated or could involve cyber operators.

The connection from the real network to the network interface defines what can be represented. However, LRM-RAG accounts for supplemental information being added to the RL agent's representation in subsequent steps. And so, concern should be directed to the ratio of the information in the RL agent's representation directly observed through the network interface to the information supplemented in various subsequent layers. There isn't a universal value judgment that can be made as to whether this ratio should be higher or lower. It should be noted, though, that a low-latency or low-information connection from the real network to the network interface necessitates a high ratio of supplemental information. It can be noted, too, that a low ratio of supplemental information suggests an implicit requirement for the attack graph generation, MDP construction, or RL agent to imbue the RL agent's representation with those concepts typically treated with supplemental information, e.g., terrain or adversary infrastructure.

The other connection is from the network interface to the real network. In LRM-RAG, the translation of the pattern of states, actions, and rewards generated by the RL agent into interactions with real networks is treated as a process internal to the network interface. This modeling decision – the decision to abstract away the interpretation of RL solutions to MDPs by path analysis programs or cyber operators – follows LRM-RAG's focus on RL. With respect to grounding, RL agents need a well-grounded means of observation as well as a well-grounded means of actuation. Significant post-processing of MDP solutions may mean greatly biasing information as it travels from the RL agent to the network. Therefore, modeling this kind of post-processing as internal to the network interface aligns with a design desiderata to maintain the RL agents two-way grounding with the real network.

5.2.2 Attack Graph Generation Processes

The next layer is termed *attack graph generation processes*. It corresponds to Kaynar's taxonomy for generating attack graphs [14]. Attack graphs are largely structural constructs that represent the topology of a network in consideration of vulnerabilities and notions of adversary success.

Definition 5.3 Reachability Analysis.
Reachability analysis discovers the basic topology of the network and produces a description of the connectivity between hosts and subnets.

Definition 5.4 Attack Graph Modeling.
Attack graph modeling selects an attack graph model, e.g., from Table 4.1, that is appropriate for the use case.

Definition 5.5 Core Building Phase.
The *core building phase* generates an attack graph corresponding to a selected model.

Kaynar can be consulted for more detail on attack graph generation [14].

5.2.3 MDP Construction Processes

The next layer is termed *MDP construction processes*. Given a graphical model of a network, this layer adds additional structure (particularly in the form of terrain) and adds behavior (in the form of transition probabilities and rewards).

Whereas the previous layer is meant to generate a generic state space for the real network, the MDP construction processes need to generate representations for the various penetration activities an RL agent is to perform. This means that there may be many specific MDPs associated with an attack graph. LRM-RAG treats the construction of these MDPs as a sequential process of layering detail on top of an attack graph.

Definition 5.6 Generic MDP.
The *generic MDP* is foundational for subsequent layers and its basic purpose is to represent the generic value (if any) of exploiting vulnerabilities and the likelihood of successfully exploiting vulnerabilities.

Using the state space provided by attack graph generation, the generic MDP initializes rewards and transition probabilities for later layers to modify. A common approach is to use vulnerabilities databases, e.g., the common vulnerability scoring system (CVSS), as a scalable approach to modeling behavior.

Definition 5.7 Terrain MDP.
The *terrain MDP* imbues the generic MDP with terrain, specifically, the kind of terrain uncovered through intelligence preparation of the battlefield.

Terrain is general, meaning adversaries and tasks on a network share similar terrain.

Definition 5.8 Adversary MDP.
The *adversary MDP* scopes the RL agent's representation by tailoring the MDP to specific adversaries.

Adversaries are modeled primarily by using attack templates, attack patterns, knowledge of red-team capabilities and infrastructure, etc. to define the set of actions an RL agent can take, but can also include varying transition probabilities with adversary skill or sophistication.

Definition 5.9 Task MDP.
The *task MDP* refines the adversary MDP to specify particular penetration testing activities.

In general, the information specifying individual RL tasks is supplemental or otherwise not provided by the real network. A common effect of the task layer is to reallocate reward or to rescope the state space.

5.2.4 Machine Learning Processes

The final layer is termed *ML processes*. At the highest level of abstraction, ML is used to approximate functions associated with penetration testing. RL is used to directly learn the MDP, while higher-level, meta-learning processes learn-to-learn.

Definition 5.10 Reinforcement Learning Agent
The *reinforcement learning agent* is an agent that maximizes the sum of discounted, expected future rewards in a MDP modeled over attack graphs.

The RL agent interacts with the task MDP, and, the outcomes of that interaction are shared with the network interface. In effect, this creates two closed loops. The first, smaller loop is that between the RL agent and the task MDP. The second, much longer loop is that between the RL agent and real network. Transfer learning, multi-task learning, and other means of augmenting RL processes that are directly used to learn the MDP are at the same level of abstraction as the RL agent. Meta-learning occurs a level higher as its own learning process and, naturally, can have its own transfer learning, etc.

Definition 5.11 Meta-Learning Agent.
The *meta-learning agent* is responsible for learning-to-learn new tasks in (new) networks.

As tasks change, e.g., over the duration of an attack campaign, and as networks change, e.g., due to a dynamic user base, new RL agents may need to be instantiated. Meta-learning takes a learning-to-learn approach to efficiently training new agents [18].

5.2.5 LRM-RAG Review

LRM-RAG is depicted in Figure 5.2 and can be summarized as follows. Information from the real network is abstracted into an attack graph by the attack graph generation processes. This attack graph is a largely structural representation of vulnerabilities, their pre- and post-conditions, and (sometimes) hosts. To formulate an environment for the RL agent, a MDP is modeled over the generated attack graph using the MDP construction processes. The result of this is a layering of behavior and additional structure on top of the attack graph. The RL agent learns by interacting with the MDP, and as tasks, adversaries, and terrain change, the RL agent can transfer learn (i.e., share knowledge) or meta-learn (i.e., learn-to-learn) and thereby reapply lessons learned across penetration testing activities. The RL agent's interaction with the MDP is interpreted by the network interface to realize penetration testing and its outcomes in reality.

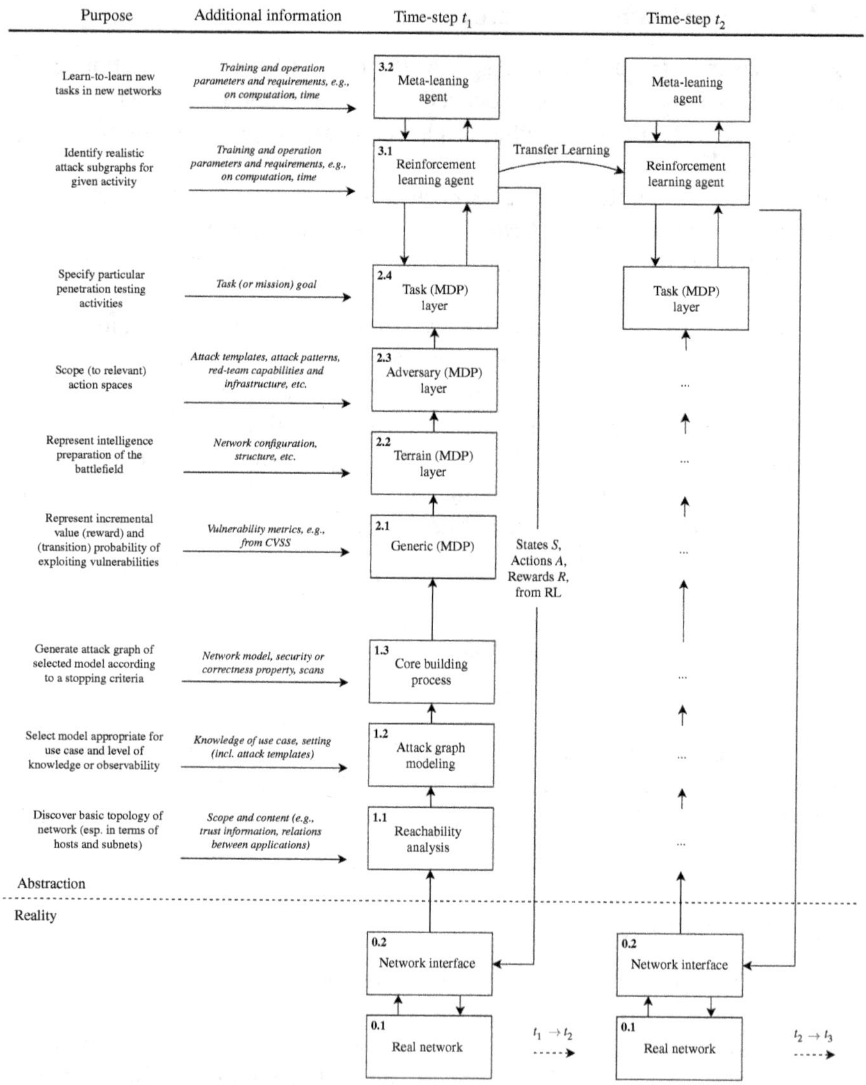

Figure 5.2 A layered reference model for automating penetration testing using RL and attack graphs. Source: From Cody [6], 2022, IEEE.

5.2.6 LRM-RAG Limitations

As stated, the LRM-RAG is a reference model – not a survey. The view offered of RL and attack graphs by LRM-RAG does not model the specifics of penetration tasks. In the case that RL is treated as automating the general task of penetration testing, those tasks are essentially specifications on the MDP, i.e., they are at a lower level of abstraction than LRM-RAG aims to model. Alternatively, in the case that RL is treated as automating specific penetration testing activities, it is not obvious what more can be generally said at a systems level about specific tasks other than the basic layering of information involved in formulating them. Similarly, LRM-RAG does not survey or explicitly define state, action, and reward spaces. Formulations of MDPs for RL with attack graphs are a varied and evolving research topic. LRM-RAG identifies an emerging, general systems structure implicit to existing formulations, as evidenced by the connections drawn between existing literature and LRM-RAG in Section 5.3.

Also, the view offered of RL and attack graphs by LRM-RAG does not model the specifics of penetration tasks. In the case that RL is treated as automating the general task of penetration testing, those tasks are essentially specifications on the MDP, i.e., they are at a lower level of abstraction than LRM-RAG aims to model. However, whole campaign emulation (WCE) fills in this gap. Alternatively, in the case that RL is treated as automating specific penetration testing activities, it is not obvious what more can be generally said at a systems level about specific tasks other than the basic layering of information involved in formulating them. Similarly, LRM-RAG does not survey or explicitly define state, action, and reward spaces. Formulations of MDPs for RL with attack graphs are a varied and evolving research topic, as discussed above.

5.3 Key Challenges for Operationalizing RL

Sections 5.3.1, 5.3.2, 5.3.3 demonstrate the use of LRM-RAG for orienting engineers to key challenges to penetration testing with RL and attack graphs. First, generation and actuation are discussed. Then, realism is related to the MDP construction processes. Lastly, challenges in unstable and evolving networks are elaborated in terms of LRM-RAG.

5.3.1 Generation and Actuation

There are several challenges to generating attack graphs for RL agents and applying their policy or related analysis to real networks.

- What is the role of supplemental information?
- How exhaustive and succinct an attack graph is needed?
- How much of the network model should be embedded into the attack graph?

LRM-RAG does not address these challenges but offers a structured framework to support discussion.

To review, the real network may be only partially observable, may be uncertain, and may be dynamic. The connection from the real network to the network interface defines what can be represented. However, LRM-RAG accounts for supplemental information added to the RL agent's representation in subsequent steps, as depicted in Figure 5.1. And so, some concern should be directed to the ratio of the information in the RL agent's representation directly observed through the network interface to the information supplemented in various subsequent layers.

Again, consider that there is not a universal value judgment that can be made as to whether this ratio should be higher or lower, but, also, consider that a low-latency or low-information connection from the real network to the network interface suggests an implicit requirement for a high ratio of supplemental information. It can be noted, too, that a low ratio of supplemental information suggests an implicit requirement for the attack graph generation, MDP construction, or RL agent to imbue the RL agent's representation with those concepts typically treated with supplemental information, e.g., terrain.

The character of supplemental information is coupled to the choice of attack graph model, and the choice of attack graph model is preceded by a choice of whether or not RL should be involved directly in the attack graph generation process. RL agents generating their own MDPs from reachability analysis may require a significant amount of representation learning to learn concepts like pre- and post-conditions, e.g., that seem essential to penetration testing. On the other hand, additional information needs to be layered into MDPs in the first place because of the narrow scope of highly optimized attack graph generation tools. These trade-offs are system, stakeholder, and use case specific.

The existing literature has taken two approaches to the generation challenge. As reported in Table 4.2, Yousefi et al. [20], Chowdhary et al. [5], Hu et al. [12], and Gangupantulu et al. [9, 10] all use MulVal, while others use some form of network model – i.e., the network model is not processed into a graph of attack paths prior to interaction with RL. Instead, RL agents use network attack simulators or hand-crafted network models. This divide seems to be tied to the role of supplemental information, as all authors using MulVal use the CVSS to provide vulnerability information.

MulVal, run to completion, will produce exhaustive and succinct attack graphs – attack graphs consisting of all and only the attack paths violating a given correctness or security property. The resulting environment $\mathcal{E}$ is therefore necessarily biased toward particular attacks and adversaries. Using network models as the environment $\mathcal{E}$ provides a kind-of generic template that can be refined using the MDP construction processes, and therefore without re-generating an (exhaustive and succinct) attack graph. While the scalability of retaining full network models, i.e., not processing them into proper attack graphs, in future networks is dubious, it is unclear if MulVal or other existing attack graph models are able to scale either.

The other connection is from the network interface to the real network. RL agents need a well-grounded means of observation as well as a well-grounded means of actuation. In LRM-RAG, the translation of the pattern of states, actions, and rewards generated by the RL agent into interactions with real networks is treated as a process internal to the network interface. This modeling decision – the decision to abstract away the interpretation of RL solutions to MDPs by path analysis programs or cyber operators – follows LRM-RAG's focus on RL.

Significant post-processing of MDP solutions may mean greatly biasing information as it travels from the RL agent to the network. The literature underreports current practice for translating the interactions between attack graphs and RL into actions in the real network, but LRM-RAG suggests two loops grounding RL agents – a smaller loop between the agent and attack graph and a larger loop between the agent and the network interface. These two loops will be discussed in more detail in Section 5.3.3.

5.3.2 Realism

Emulating realistic adversary behavior in networks is another key challenge to using RL with attack graphs. There are countless examples of RL agents solving MDPs by finding software bugs or modeling errors in their environment $\mathcal{E}$ which generate reward but do not lead the agent to completing its desired task. In RL with attack graphs, allotting reward for traversing a network switch undetected, e.g., could create positive feedback that incentivizes the agent to continuously loop through already exploited hosts to reuse the network switch. For this reason, reward engineering is a difficult and avoided task in RL.

LRM-RAG breaks MDP engineering into four parts. The generic MDP uses the CVSS to furnish incremental rewards for exploiting vulnerabilities and assigning transition probabilities [5, 10, 12, 20]. Nearly half the authors in Table 4.2 implicitly use a generic MDP, however, even those that use it remark on its deficiencies. The terrain MDP addresses these deficiencies by introducing general concepts of terrain which apply to adversaries and tasks generally. Gangupantulu et al. [9, 10] and Cody et al. [7] propose several terrain concepts for RL with attack graphs including obstacles, key terrain, and cover and concealment.

Terrain may give a realistic network behavior, perhaps, but terrain is not specific enough to give realistic adversary behavior. Adversaries have some preference over terrain, meaning different adversaries with different infrastructure, etc., will behave differently with the same terrain constraints. A nuanced challenge to modeling adversaries with MDPs is that campaigns are time-dependent. MDP engineering to represent an adversary's time-varying goals can treated using the task MDP. For example, the adversary MDP can set the action space based on an understanding of adversary capabilities, and then the task MDP can scope to a specific part of the network, modify terrain penalties, and set the terminal state [7, 9].

To an RL researcher, LRM-RAG appears to suggest extensive reward engineering. The convention in the RL literature is that reward engineering bottlenecks RL performance. However, there are a number of unavoidable steps to specifying penetration testing activities. These steps certainly lead to an MDP with engineered states, transitions, and rewards. And for rewards in particular, it is highly unlikely that the full scope of detail that a penetration testing system may be tasked with emulating can be summarized by allotting large, sparse terminal rewards to agents. Thus, LRM-RAG does not suggest reward engineering, but rather identifies a natural bias toward RL solutions that rely on extensive reward (and MDP) engineering. It is unclear that this natural bias can be removed without removing the meaning of penetration testing.

5.3.3 Unstable and Evolving Networks

Recall the basic structure of LRM-RAG. Information from the real network is abstracted into an attack graph by the attack graph generation processes. The generated attack graph is a largely structural representation of vulnerabilities, their pre- and post-conditions, and (sometimes) hosts. To formulate an environment for the RL agent, a MDP is modeled over the generated attack graph using the MDP construction processes. The result of this is a layering of behavior and additional structure on top of the attack graph. The RL agent learns by interacting with the MDP, and as tasks, adversaries, and terrain change, the RL agent can transfer learn (i.e., share knowledge) or meta-learn (i.e., learn-to-learn) and thereby reapply lessons learned across penetration testing activities. The RL agent's interaction with the MDP is interpreted by the network interface to realize penetration testing and its outcomes in reality.

The long chain of functions between RL agents and real networks *is* an RL agents grounding to reality. When information in an RL agent's representation is made available primarily by lower-level processes, the RL agent's representation is more directly grounded to the real network than when a majority of information comes from higher-level processes, e.g., from reward engineering tasks and terrain or from hand-engineering action sequences. But there is a dynamic aspect to the grounding problem as well.

When the real network undergoes a change, that change is communicated to the ML processes by way of the network interface, attack graph generation processes, and MDP construction processes. Clearly, if the rate at which the real network changes is faster than the rate at which the RL agent's MDP can be created, then the RL agent will be grounded to an out-of-date representation of the real network while such a change occurs (and for a period thereafter). For largely static enterprise networks, when analysis is offline or when changes are small, this may not be a large concern. But in dynamic networks, a careful study should be made.

Recall the two loops identified in Section 5.3.1. In the smaller loop, the RL agent searches over attack graphs by way of interaction to find a particular subgraph or path. In the larger loop, this search process or discovered subgraph informs changes to the real network, e.g., it informs network hardening or resilience measures. In theory, these two loops could be similar in length if, e.g., the RL agent took actions in the attack graph at the same rate that the attack graph is refreshed. However, since once the task MDP is formed, its analysis is entirely computational, RL agent interaction with the MDP can likely, in general, be at least as fast. For example, the RL agent could interact every millisecond with an attack graph that is refreshed every 30 s.

Such a rate differential may make minute-to-minute host variations, e.g., due to virtual private network (VPN) use, a nonissue. But, in future networks, e.g., software-defined networks and Internet of Things networks [2, 3], the rate at which networks can evolve will seriously challenge automated penetration testing systems [4]. Moreover, emulating phenomena like payload mutation and intelligent entry-point crawling may require learning under the assumption that networks are unstable and evolving, i.e., RL will be tasked with emulating adversary responses in changing networks.

5.4 Conclusions

RL-based attack graph generation has the potential to provide continuous cyber capabilities like test and evaluation (T&E) to organizations from the scale of small businesses to nation-states. The literature is quickly being populated with methods for emulating specific stages of attack campaigns, but, these methods fall short of end-to-end emulation of attack campaigns. From Chapter 4, WCE provides an integrative concept and future research direction for automated penetration testing with RL and attack graphs. Importantly, WCE places a focus on MDP engineering, identifying engineering MDPs for specific stages and engineering their composition as the primary challenges to advancing automated penetration testing with RL and attack graphs. Here, in Chapter 5, LRM-RAG provides an architecture for operationalizing a layered, MDP engineering process for WCE. While RL algorithms are an important area of research, well-established MDP construction methods for RL-based generation of attack graphs do not yet exist. Thus, it is unclear how to benchmark RL algorithm research for attack graph generation, and unclear what, if anything, state-of-the-art RL algorithms are missing.

Also, the task of grounding RL agents to real networks is an open area of research. The choice to use attack graphs is not only a choice of (network) representation (for the RL agent), but also it is a specification on the grounding mechanism. Attack graphs are abstractions that must be generated, and that generation takes time. Consider, attack graphs (currently) grow at best log-linear and at worst exponentially with the number of hosts in the real network. Efforts to make attack graph generation computationally efficient highly bias representations, e.g., by not representing hosts, in effect, trading off the latency of the grounding mechanism for resolution in (network) representation.

The information needed but left out by the generation process must be re-introduced. Customizable representations that can fit models of terrain or adversaries are highly desired in penetration testing activities, but relying on external information makes grounding tenuous. When a network destabilizes, changes, or otherwise evolves, detailed modeling becomes a bottleneck in re-establishing an RL agent's grounding. To that end, dynamic networks prefer automated and empirical solution methods to constructing representations. In contrast, translating an RL agent's states and actions into interactions with real networks gives a nearly direct grounding. This asymmetry underpins LRM-RAG and is characteristic of its framing of continuous penetration testing with RL agents and attack graphs.

LRM-RAG provides a systems-level, extensible reference model for engineering the grounding of RL agents and their attack graphs to real networks. Domain experts can use LRM-RAG to give context to the individual engineering problems with which they are faced: attack graph generation, MDP formulation, RL algorithm design, etc. And, naturally, systems engineers can use LRM-RAG to relate those problems to each other.

The next chapter, "Toward Practical RL for Pen-Testing," explores taking the operationalized RL techniques and applying them to real-life cybersecurity scenarios.

References

1 Mridul S Barik, Anirban Sengupta, and Chandan Mazumdar. Attack graph generation and analysis techniques. *Defence Science Journal*, 66(6):559, 2016.

2 Wiem Bekri, Rihab Jmal, and Lamia C Fourati. Softwarized internet of things network monitoring. *IEEE Systems Journal*, 15(1):826–834, 2020.

3 Oladayo Bello and Sherali Zeadally. Intelligent device-to-device communication in the internet of things. *IEEE Systems Journal*, 10(3):1172–1182, 2014.

4 Chung-Kuan Chen, Zhi-Kai Zhang, Shan-Hsin Lee, and Shiuhpyng Shieh. Penetration testing in the IoT age. *Computer*, 51(4):82–85, 2018.

5 Ankur Chowdhary, Dijiang Huang, Jayasurya S Mahendran, Daniel Romo, Yuli Deng, and Abdulhakim Sabur. Autonomous security analysis and penetration testing. In *2020 16th International Conference on Mobility, Sensing and Networking (MSN)*, pages 508–515. IEEE, 2020.

6 Tyler Cody. A layered reference model for penetration testing with reinforcement learning and attack graphs. In *2022 IEEE 29th Annual Software Technology Conference (STC)*, pages 41–50. IEEE, 2022.

7 Tyler Cody, Abdul Rahman, Christopher Redino, Lanxiao Huang, Ryan Clark, Akshay Kakkar, Deepak Kushwaha, Paul Park, Peter Beling, and Edward Bowen. Discovering exfiltration paths using reinforcement learning with attack graphs. *arXiv preprint arXiv:2201.12416*, 2022.

8 Greg Conti and David Raymond. *On cyber: towards an operational art for cyber conflict*. Kopidion Press, 2018.

9 Rohit Gangupantulu, Tyler Cody, Abdul Rahman, Christopher Redino, Ryan Clark, and Paul Park. Crown jewels analysis using reinforcement learning with attack graphs. In *2021 IEEE Symposium Series on Computational Intelligence (SSCI)*, pages 1–6, 2021.

10 Rohit Gangupantulu, Tyler Cody, Paul Park, Abdul Rahman, Logan Eisenbeiser, Dan Radke, and Ryan Clark. Using cyber terrain in reinforcement learning for penetration testing. In *2022 IEEE International Conference on Omni-layer Intelligent Systems (COINS)*, pages 1–8, 2022.

11 Stevan Harnad. The symbol grounding problem. *Physica D: Nonlinear Phenomena*, 42(1–3):335–346, 1990.

12 Zhenguo Hu, Razvan Beuran, and Yasuo Tan. Automated penetration testing using deep reinforcement learning. In *2020 IEEE European Symposium on Security and Privacy Workshops (EuroS&PW)*, pages 2–10. IEEE, 2020.

13 Somesh Jha, Oleg Sheyner, and Jeannette Wing. Two formal analyses of attack graphs. In *Proceedings 15th IEEE Computer Security Foundations Workshop. CSFW-15*, pages 49–63. IEEE, 2002.

14 Kerem Kaynar. A taxonomy for attack graph generation and usage in network security. *Journal of Information Security and Applications*, 29:27–56, 2016.

15 Xinming Ou, Wayne F Boyer, and Miles A McQueen. A scalable approach to attack graph generation. In *Proceedings of the 13th ACM Conference on Computer and Communications Security*, pages 336–345, 2006.

16 Francesca Poggiolesi. Grounding principles for (relevant) implication. *Synthese*, 198(8):7351–7376, 2021.

17 Oleg Sheyner, Joshua Haines, Somesh Jha, Richard Lippmann, and Jeannette M Wing. Automated generation and analysis of attack graphs. In *Proceedings 2002 IEEE Symposium on Security and Privacy*, pages 273–284. IEEE, 2002.

18 Joaquin Vanschoren. Meta-learning: a survey. *arXiv preprint arXiv:1810.03548*, 2018.

19 Vladimir Vapnik. *The nature of statistical learning theory*. Springer Science & Business Media, 1999.

20 Mehdi Yousefi, Nhamo Mtetwa, Yan Zhang, and Huaglory Tianfield. A reinforcement learning approach for attack graph analysis. In *2018 17th IEEE International Conference on Trust, Security and Privacy in Computing and Communications/12th IEEE International Conference on Big Data Science and Engineering (TrustCom/BigDataSE)*, pages 212–217. IEEE, 2018.

21 Jianping Zeng, Shuang Wu, Yanyu Chen, Rui Zeng, and Chengrong Wu. Survey of attack graph analysis methods from the perspective of data and knowledge processing. *Security and Communication Networks*, 2019:1, 2019.

6

Toward Practical RL for Pen-Testing

6.1 Current Challenges to Practicality

6.1.1 The Problem of Scaling

Simulations of a network obviously must scale to the size of the network in question, and for real enterprise networks, this can be quite large (10s of thousands of hosts). Scaling for a machine learning model can mean a number of things. Poor scaling, could mean that it is simply computationally expensive, in resources or in time, to train the model or run predictions. If such a model is meant to save time and money in real red teaming exercises, then the computational cost and run time have hard limits before it is simply not worth it to run the simulation versus a human team that could perform the same task in less time and at a lower cost. However, when concerning RL with large state and action spaces, the consequences of scaling are often much more dramatic, in that the model simply won't converge. A stable policy will not be learned, and the training exercise simply will not finish. This could mean that no training epochs reached the stopping criteria (such as privilege escalation on a particular host), or even if episodes are completed, the reward may simply continue to bounce around wildly indefinitely (Figure 6.1).

Early attempts at applying RL to penetration testing similar to the methodology in this text have been extremely limited in the scope of their tests, often limited to networks of less than 10 machines [15]. Intelligent automated penetration testing system (IAPTS) (2018) tested on a network of just 10–100 machines [5, 6]. Research efforts that attempt tests with larger scale networks (100s of hosts) are unrealistically limited in other ways, such as having extremely limited action spaces [1, 2]. Many of these utilize deep Q learning.

Reinforcement Learning for Cyber Operations: Applications of Artificial Intelligence for Penetration Testing, First Edition. Abdul Rahman, Christopher Redino, Dhruv Nandakumar, Tyler Cody, Sachin Shetty, and Dan Radke.

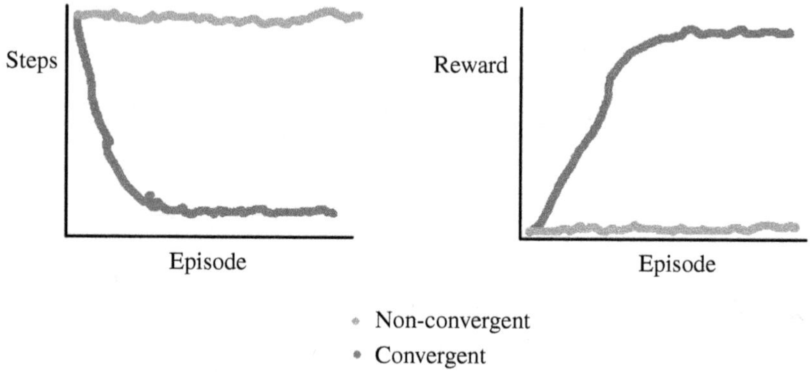

Figure 6.1 Convergent vs. non-convergent learning. When a model does not converge on an effective policy, the number of steps per episode will always bounce around some high value. The reward bounces around some noisy floor value for a non convergent model, when the agent never completes its mission, and plateaus around an optimal value for convergent agents that complete their mission goals.

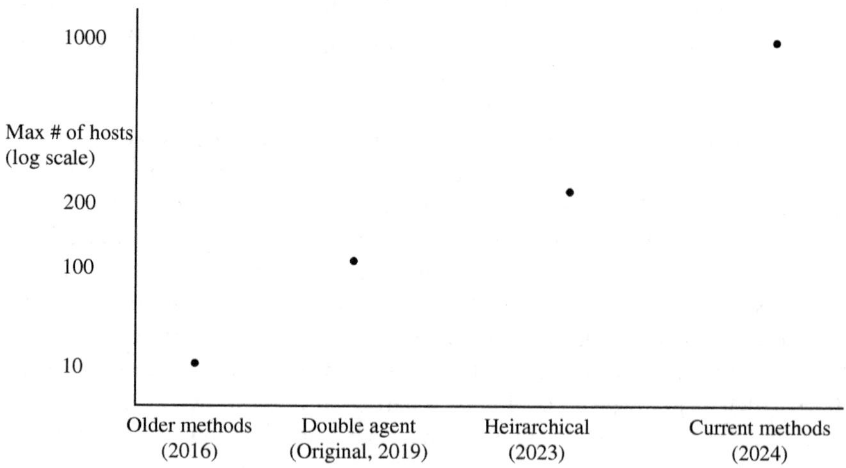

Figure 6.2 The rough maximum number of hosts an RL methodology for cyberattack simulation has successfully scaled to over time. These limits often indicate the maximum size of network for which a model can even converge on policy and are independent of how fast a policy can be learned. Only recently are methodologies for realistically sized networks even feasible.

It is no longer a question of diminishing returns with scaling but a binary success or failure based on network size. In this way, each well-defined methodology for RL, when applied to penetration testing, can be thought of as having an approximate cap on the size of the network it can be applied to (Figure 6.2).

Even before applying fancy algorithms or architectures to address scaling, the scalability of any RL methodology begins with how the state and action spaces are defined, that is, how we define the MDP.

The previous chapter discussed how the relationship between an attack graph and MDP is not necessarily straightforward, and that there are choices to be made in designing the state space so that it is scalable. The LRM-RAG framework starts from a subset of the total graph, and then the task of the agent is to find the subgraph that is a realistic attack path through the network. The work in [3] where LRM-RAG was first proposed goes into more depth here to compare how some different approaches build attack graphs, and how the size of the graph grows with the number of hosts. This, in part, depends on what an actual node in the graph represents, as it is not always directly a host, but it can itself be a state, encoding the current status of the host based on the attack path leading to it.

The model framework described by LRM-RAG is a useful reference model, and the methodology described here will borrow from it again later in this chapter. But for the moment, let's imagine moving away from the concept of the attack graph explicitly, and think on how a real attacker has access to information and how they act on it. In the RL frameworks described here, the "attacker" is synonymous with the agent in the model, but it is important to call out the distinction of information available to the agent and information available to the model as a whole. The challenges in scaling attack graphs described in [3] are made even more difficult when considering partial information. This is not just an academic consideration, but rather one grounded in reality. In real attacks, an actual adversary is always operating on partial information. In real red teaming exercises, even those that are more "white box," scan information is often incomplete or out of date, and there's an unavoidable element of "learning as you go," which sounds quite a bit like RL. Attack graphs accommodate partial information by growing even larger, with new branches to encode the different possibilities obscured by the missing information. Instead of formulating the problem as a top-down approach (start with a larger attack graph and whittle down to a path), the approach is more akin to what a human attacker would be a bottom-up approach, starting with almost no information, scan from where you have a foothold and piece by piece built out the attack path a single hop, a single action at a time.

The reality is, that real human brains do a combination of the top-down and bottom-up approach, they unlikely have a full attack graph in mind, (unless it's a very small and trivial network), but they have done some degree of preparation, recon, and even if the direct information here is limited, their intuition, their experience guides the way as they gather more information. In later sections, we will describe an RL framework inspired by this that ideally can be both scalable and realistic.

This problem of scaling in RL is not unique to penetration testing or cyber applications, this difficulty comes anytime there is a very large action space, and there are some common strategies in the wider RL literature for trying to deal with this.

As already stated above, the very definition of the state action space can have a profound impact on the scalability of the solution, and whereas we will attempt to put forth a scalable model based on analogies to real attackers, a common theme in current literature is to simplify the action space. Simply having a smaller set of options for an agent to choose from makes learning, convergence, and performance easier. This could be a small, but realistic subset of actual exploits that could be used, as in [14]. Other works, such as [16], make use of generalized actions that can account for many similar ones, such as a generic exploit action or a generic scanning action. By rolling up multiple actions into a single smaller set, there are economic savings for the model, a smaller simpler space for the agent to explore. Even in the case of [16] where there is a single generic type of exploit, much of the current literature often explicitly represents the action space as the number of operations (exploits, scans, etc.) times the number of nodes, and so for realistically sized commercial networks, the action space can still be vast.

Simplifying the action space in RL for penetration testing can be crucial to making the learning process more efficient and effective. There are a variety of strategies to simplify the action space in the context of penetration testing, and we will review some of the most common ones here. A simple way to make the problem easier for the agent is to reduce the dimensionality of the problem. Focus on a subset of relevant actions: Identify the most critical actions that an attacker might take during penetration testing. Limit the action space to a subset that captures the essential steps in an attack. Combine similar actions: If there are actions that are conceptually similar, consider merging them into a single action to reduce the dimensionality of the action space. In a similar vein, one could focus on specific attack scenarios; instead of considering the entire range of possible actions, narrow the focus to specific attack scenarios or types of vulnerabilities. This can help in tailoring the action space to the specific goals of the penetration test. This can be taken further by breaking specific attack campaigns into parts, such as recon, infiltration, and exfiltration. An extreme version of this would be to use a predefined playbook. Define a set of predefined attack playbooks; limit the actions to a set of predefined attack scenarios or playbooks commonly used in penetration testing. This can simplify the action space and guide the RL agent towards learning more practical and effective strategies.

A little more sophisticated than just sub-setting is to make abstractions of the action and group them into categories. Instead of considering each individual action separately, group related actions into categories. For example, actions related to reconnaissance or exploitation could be abstracted into broader categories. If we want the reduced action space of using abstracted categories but

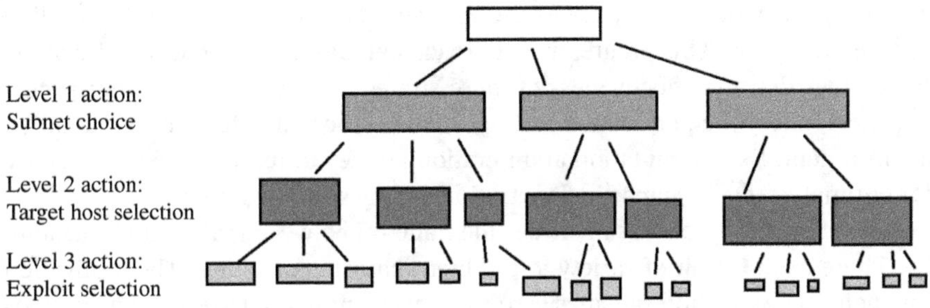

Figure 6.3 A visualization of a hierarchical action space. At the top level, an agent only has to decide between a few options and the with subsequent actions the agent moves down the diagram, never having too many options presented to it at once.

don't want to oversimplify what is actually done by the agent, we may consider a hierarchical action space (Figure 6.3).

To use a hierarchical RL approach, we would organize actions hierarchically, where high-level actions represent overarching strategies and low-level actions correspond to specific steps within each strategy. This can simplify the learning process by allowing the RL agent to focus on learning high-level decision-making first. An example of a hierarchical action space for penetration testing in practice is described in [9]. A thoughtful prioritization of actions can be helpful. To assign priorities to actions, one would identify the most critical actions and assign higher priorities to them. This helps the RL agent concentrate on learning the most impactful steps in a penetration test. Of course, this can introduce a bias in the model and does a bit to presuppose how the agent should behave. An indirect way to reduce the action space is to reduce the state space. By identifying key features or indicators that capture the state of the environment and focusing on critical features, you can reduce the complexity of the state space and, consequently, the action space. In some RL applications (and in certain attack scenarios), the discretization of continuous actions can effectively reduce the action space. If the original action space is continuous, consider discretizing it into a finite set of discrete actions. This simplification can make the learning process more tractable. While it would seem at first glance that most actions in a penetration test are already discrete, a couple of notable exceptions are any sort of wait action (which can be any arbitrary amount of wall-clock time) and moving of payload, which could be (nearly) any amount of data at a time. A key factor often undersold in the literature is domain knowledge integration. Leverage the expertise of penetration testers to guide the simplification of the action space: experts can provide valuable insights into the most relevant and impactful actions. When actions are unavoidably high dimensional and complex, some form of curriculum learning can be helpful. Gradually introduce complexity: Start with a simplified action space and gradually introduce complexity as the RL agent becomes proficient in handling simpler scenarios. This progressive approach can aid in more effective learning. Remember that the effectiveness of these strategies may depend on the specific requirements of the penetration testing task and the characteristics of the target environment. Experimentation and iteration are key to refining the action space for optimal RL performance in penetration testing scenarios.

Most of these approaches involve some trade-off between realism and tractability. There can be a bit of variety to the hierarchical approaches. The motivation for such decomposition schemes is so that the agent doesn't see as many options simultaneously at any one time. The agent could pick from a class of actions (say the family of all exploits) and then later pick the specific exploit, or perhaps the agent first makes a decision on which subnet to perform an action in and then selects from actions within that subnet. These types of approaches have been proposed as far back as [11]. This type of decomposition naturally lends itself to schemes with multiple agents working together, each learning how to deal with its own action space. The use of multiple agents in RL can offer several benefits, providing advantages that may not be achievable with a single-agent approach. Some key benefits of employing multiple agents in RL include:

- **Diverse Exploration:** Multiple agents can explore the environment in diverse ways, discovering a wide range of strategies and solutions. This diversity enhances the exploration-exploitation trade-off, leading to a more comprehensive understanding of the state space.
- **Knowledge Sharing:** Agents can share experiences and insights, facilitating collective learning. This knowledge-sharing mechanism allows the system to benefit from the expertise gained by individual agents, resulting in faster learning and improved overall performance.
- **Specialization:** Agents can specialize in different aspects of the task or environment. Specialization allows each agent to focus on specific subproblems, leading to more efficient and effective solutions for different facets of the overall problem.
- **Parallelization**: Training multiple agents in parallel can significantly accelerate the learning process. This is particularly advantageous in situations where simulations or computational resources are available, as parallelization allows for simultaneous exploration of different parts of the state space.
- **Competitive Dynamics**: Introducing competition among agents can foster the development of robust strategies. Competitive dynamics encourage agents to continuously improve their policies to outperform others, resulting in the emergence of sophisticated and resilient behaviors.
- **Transfer Learning**: Agents can transfer knowledge gained in one scenario to another. This transfer learning capability enables the system to leverage

insights from previous experiences, adapting more quickly to new tasks or environments.

- **Adversarial Training**: Multiple agents can simulate adversarial scenarios, helping the system learn to defend against a variety of attacks. Adversarial training enhances the robustness of the system by exposing it to diverse challenges and threats.
- **Coordinated Decision-Making:** Agents can coordinate their actions to achieve common goals. Coordinated decision-making is beneficial in tasks that require collaboration or teamwork, as agents can work together to solve complex problems.
- **Ensemble Learning**: Combining the policies learned by multiple agents into an ensemble can result in a more robust and generalizable policy. Ensemble learning mitigates individual agent biases and errors, leading to a more reliable system.
- **Social Learning Dynamics**: Interaction among multiple agents can lead to the emergence of novel and effective behaviors. Social learning dynamics enable the system to adapt based on the collective behavior of the agents, fostering the development of intelligent strategies.
- **Improved Exploration-Exploitation Balance**: The presence of multiple agents can naturally balance the exploration-exploitation trade-off. While some agents explore new possibilities, others may exploit known strategies, ensuring a more well-rounded approach to learning and decision-making.

It's important to note that the benefits of using multiple agents depend on the specific characteristics of the problem domain and the design of the multi-agent system. Careful consideration of communication, coordination, and learning mechanisms among agents is crucial for realizing these advantages effectively. In a later section, when we propose our own approach, we will take inspiration from all of these while trying to maintain realism.

6.1.2 The Problem of Realism

Perhaps less obvious than the issue of scalability, simulations of penetration testing can result in unrealistic results in varying degrees. The attack path output by a simulation might not actually work in practice on the actual live network, due to bad assumptions of the model or outdated scan information. Even if the attack path is technically valid and possible, in practice, the actions taken by the agent, in the order they were taken and with the timing they were taken, would most likely have alerted some defense to the malicious behavior, compromising the intent of the mission. Even if the path is valid and sufficiently stealthy, it may yet still be a very **unnatural** path that a human attacker would never take. This last point is particularly interesting in that it highlights the difficulty of the problem here. The solution space in real networks at scale is usually actually vast. A real attacker may exploit a series of seemingly unrelated, noncritical vulnerabilities, that in concert achieve a larger, more malicious goal [5, 6]. When described in terms of game theory, cybersecurity experts will say that the game is played nowhere near optimally from either side (offensively or defensively). This inefficient, nonoptimal scenario has practical implications for how to play both offense and defense (and thus how to simulate these as well). The defense is like a dam, holding back water, and the vulnerabilities are cracks in this dam, which, with pressure from an attack, can rip the dam open and cause a flood. The inefficiency of the defense is that there is no hope to caulk over every crack: systems are too intricate, layered, and distributed. To make things even more difficult, it is often challenging for network owners to even know what is on their own network. Security professionals don't have the time or resources to find and fill every hole, especially not in a way that would still make the network functional for its intended purpose, and good defensive strategy at best hopes to fill the most likely targeted holes and know ahead of time where to monitor and respond when an incident does occur. Developing this strategy of filling and monitoring is what pen-testing is intended to do. The offense, meanwhile, is also inefficient, and like the defense, it will not find every hole either, for similar reasons, limited information, time, and resources. It then becomes a very asymmetric game; the offense wants to make its move where it's not likely to be guarded, and the defense wants to anticipate this move. Good defense, therefore is about anticipation, not exhaustion. A model that produces just a few very likely attack paths has more value than a model that can produce dozens of attack paths that would never be exploited. Therefore, a good RL simulation of pen-testing should attempt to capture the motivation of an attack. The actual actions and strategy are what the model is supposed to learn on its own (the policy), but the motivation can be encoded into the reward. A model with the reward engineered in an unnatural way will get the motivation and likely the behavior wrong, producing unrealistic (if not infeasible) attack paths. The interesting wrinkle to all this is that operators are an inventive bunch.

A frequent practice in current RL for pen-testing research is to build rewards directly out of the common vulnerability scoring system (CVSS), for example, in [14]. This is a well-motivated practice in that CVSS provides a standardized, quantitative, objective, methodology to measure vulnerabilities present in a network. Relying solely on agents trained using the CVSS fails to accurately depict reality. This is because the generated attack graphs lack the operational nuances found in intelligence preparation of the battlefield (IPB), which incorporates concepts of (cyber) terrain [17]. In simpler terms, a strategy based solely on assessing vulnerability severity lacks a comprehensive understanding of the broader context. Human red teams and adversaries consider a wealth of information, viewing the entire landscape rather than fixating on individual vulnerabilities, an aspect overlooked by approaches focused solely on CVSS. While studies by Yousefi et al., Chowdary et al., and Hu et al. leverage CVSS for constructing attack graphs, similar to the approach taken by Gallon and Bascou, it is important to note that CVSS scores, while an industry-standard metric, do not always provide a meaningful contextual perspective for cyber operators. Relying solely on CVSS abstractions in network representations can introduce bias, emphasizing vulnerabilities over a realistic understanding of how adversaries plan and execute attack campaigns. Consequently, this can result in RL methods converging to unrealistic attack scenarios.

The CVSS is a widely used framework for assessing the severity of cybersecurity vulnerabilities. However, it has several limitations that are important to consider:

- **Focus on Technical Severity:** CVSS primarily focuses on the technical aspects of vulnerabilities, such as how easily they can be exploited and the potential impact on the system. It does not take into account the broader context of the organization's security posture, including factors like the value of the targeted data, existing security controls, or the motivation and capability of potential attackers.
- **Lack of Context:** CVSS scores may not always provide a meaningful contextual picture for cyber operators. It doesn't consider the specific environment or the potential consequences of a successful exploit in a given organizational context. As a result, a high CVSS score doesn't necessarily mean the vulnerability poses a critical risk to every organization.
- **Static Scoring:** CVSS scores are assigned based on a set of criteria at a specific point in time. They do not dynamically adapt to changes in the threat landscape or the organization's security measures. This can result in scores that do not accurately reflect the evolving nature of cyber threats.
- **Solely Technical Metrics:** CVSS relies heavily on technical metrics and may not adequately address vulnerabilities related to human factors, such as social engineering or insider threats. Cybersecurity is a complex field that involves both technical and human elements, and CVSS might not capture the full spectrum of risks.
- **Difficulty in Scoring Complex Vulnerabilities:** Some vulnerabilities are complex and multifaceted, making it challenging to assign a precise CVSS score. Additionally, there can be subjective interpretations in scoring, leading to inconsistencies in assessments.
- **Limited Coverage of Exploitability:** While CVSS considers the exploitability of vulnerabilities, it may not comprehensively cover the real-world likelihood of exploitation. Factors such as the existence of known exploits in the wild and the ease of creating an exploit are not explicitly addressed.

- **Overemphasis on Specific Metrics:** The use of a formulaic approach to calculate scores may result in an overemphasis on certain metrics, potentially overlooking important aspects of a vulnerability. For example, a vulnerability with a low CVSS score might still pose a significant threat depending on the organization's context.
- **Not Suitable for All Use Cases:** CVSS was initially designed for risk managers and security experts. While it provides valuable information for these audiences, its complexity may make it less accessible for nonexperts or for organizations with limited resources for cybersecurity expertise.

Despite these limitations, CVSS remains a valuable tool for assessing and prioritizing vulnerabilities. It is important, however, to use CVSS scores as part of a broader risk management strategy that incorporates additional contextual information and considers the specific needs and characteristics of the organization.

While CVSS scores are useful in practice and currently considered an industry standard, it is important to remember that a measure of threat severity is not the same as a measure of risk and that they do not generalize to give information that's useful for evaluating an entire attack path through a network. From the perspective of an attacker, a greater risk means a greater chance of detection. While the CVSS scores of vulnerabilities do inform the probability of success of any particular exploit in the models here, the real driving force in the agent's behavior comes from reward engineering centered around concepts of terrain. The important key here is context; a real attacker considers contextual information, and their decisions are made in a holistic and eclectic manner (they use whatever they have available to them), so realistic simulation will have to capture this aspect of the attacker's nature which lies in the reward engineering.

6.2 Practical Scalability in RL

Previously, we reviewed several current methods for tackling scalability issues in RL, but most of these are not unique to penetration testing. Of key importance here though, is we don't want any decision we make for the sake of scalability to sacrifice realism, rather we will try to find an approach where we can have our cake and eat it too, where decisions for the sake of realism **also** benefit scalability. Furthermore, as detailed earlier, the outcome of scalability efforts is often binary; rather than measuring ourselves in terms of training time, we want to look at the number of network sizes that are feasible for a given model. Whereas much of the current literature caps out at 10s or 100s of nodes, the methodology described here, and shared by Cody et al. [4], Huang et al. [7], and Wang et al. [13], has been applied to networks of thousands of nodes. While this is currently impressive, it is by no means believed to be the definitive approach, even if it is a current leader. This approach did not spring out from a vacuum, but rather it builds off of many of the methodologies discussed elsewhere in this chapter and in this text. The importance of this approach to the reader should be conceptual, as a reference model to build off of to a further degree, akin to LRM-RAG.

6.2.1 State and Action Spaces

In designing the state and action spaces, to begin with, we will move away from attack graphs, though information collected in preparation for training the model will be largely the same (operating systems, ports, and services of hosts, including firewall information and enough bread crumbs to infer the topology of how these nodes are connected). The attack graph itself, however, will not be computed explicitly in this approach, though it will be replaced with a simpler proxy. Without this computation, we have tremendous savings in run time, but to repeat, the primary issue was convergence, not compute time, and it's important to reflect on what we give up by giving up attack graphs. The attack graph lays out what is **possible** so that the top-down approach from its bird's eye view is realistic. Shattering this graph, left with disparate nodes and services, the view from an attacker (agent) upon getting a foothold in the network is a seemingly insurmountable mess of possible actions, the exact situation we're looking to avoid. So how, if not with an attack graph, do we organize the state and action spaces? We take inspiration from analogy to what an actual attacker sees, and what they care about as they are making decisions.

The state of the entire network can be thought of as a giant nested vector, encoding the current status of each host.

We are only concerned with encoding those aspects of the network that change as the agent performs its attack, and here, it becomes important to remember there is a difference between what information is available to the agent and what information is available to the model as a whole. As an example, the model as a whole will have knowledge of services available on every host (assuming thorough and up-to-date scanning in preparation for model training), but the **agent** will only know the services available on those hosts, which have already performed a scanning action. Another example is that the model as a whole has information about the overall topology of the network and how every host is connected by services, whereas the agent only knows if a particular host is currently accessible or not. In this bottom-up approach, instead of whittling down to a sub-graph from a larger attack graph, we build the graph piece by piece. The bits in the states vector that get flipped as the agent performs actions should all be of the form that they answer the agent's question: "OK, what can I do next?" While we show an example vector above, this is by no means exhaustive in the works of [4, 7, 13], the state space remains thematically similar to the particulars change. This isn't just because of progress in making the methodology more realistic and robust, but rather, these different works are attempting to solve different problems with the RL, simulating different types of behavior, and the particulars of those behaviors require different information in the state space. For example, the current size of the remaining payload to be sent is relevant in simulations of exfiltration but probably not in simulations of reconnaissance. Besides information directly related to the agent's actions, such as discoverability and access level, secondary effects that are a consequence of the agent's actions are also encoded here, such as a defensive response from the network. It is important to remember that the simulation discussed here is entirely offline, so any defensive response is modeled. This modeling can be quite simple, and we are only concerned with those effects that impact the agent's future options, such as isolating a host.

With an understanding of the nature of state space (a set of toggles indicating the status of each host), the set of actions available to the agent is then all of the options the agent has to flip these toggles, to make a host discoverable, to gain access, to escalate privileges, to move payload, etc. Without care, this list of actions can quickly become quite large, and while simplifying and consolidating actions is tempting, we want to maintain realism. A compromise we make here is to simplify and consolidate actions only where it has no practical effect on the outcome. For example, perhaps for a particular host, running one particular service, running a certain version of OS, there are three known exploits that an attacker could make use of to gain root access. Our rule of thumb here would say to consolidate those three exploits to one action. We make this consolidation because we assume it makes no practical difference on the outcome of the simulation, but it is important to remember that it **is** an assumption, and with more information, we may believe these three exploits to be practically equal: are they equally stealthy, can they performed in the same amount of time, do they have the same success rate, or do some options have a probabilistic chance of failure. The amount of information available to the model builders (you) must dictate when it is appropriate to consolidate these exploits.

This guidance on when to consolidate actions is useful, but it doesn't go nearly far enough to solve the problem. Once we have a set of such consolidated actions, the idea is each one would be available to the agent anywhere it sees that particular combination of OS and service. All actions that have zero chance of success are masked such that only viable options are available at each step. This greatly reduces the action space by simply throwing out impossible options and this is feasible so long as actions are defined by their requirements. Building the actions of their requirements not only makes this masking process feasible, but it also makes action consolidation a much more objective exercise and publically available resources such as CVE, common weakness enumeration (CWE), and CVSS can make it a programmatic one.

Even with this requirements-driven action masking and consolidation, it is **still** not enough to tame the action space for large networks, so going back to the list of commonly leveraged methods, let's pick another one that works with what we've already built and is still faithful to realism.

6.2.2 A Flavor of Double Agent

Decomposition of the action space and the use of multiple agents are not new tricks, as already mentioned. Again, these are not methods used only for RL in penetration testing (PT), but are more broader, and when applying it to PT, our guiding principle will continue to be realism. If the action delegation among agents is conceptually sound from an attacker's perspective, it is more promising. Furthermore, we want a decomposition of actions that is compatible with requirements-driven consolidated actions, as in the previous section. The best candidate we've seen for this is the basic methodology laid out by Nguyen et al. [10] in what they call the double agent architecture (DAA). In this methodology, there are two principal agents: an exploration agent and an exploitation agent. The former is responsible for selecting which host to perform an action on and the other is responsible for deciding which action to perform on that host. In this manner, the selection process of the exploration host is itself a new set of actions, and the exploitation host can perform a set of actions from the consolidated requirements masked list that is appropriate for that particular host.

Traditionally, Deep Q-Network (DQN) has been the foundational algorithm in the realm of modeling RL for penetration testing. Nguyen et al. proposed an approach that employs two agent to critic agents: a structural agent, responsible for learning structural information about subnets, hosts, firewalls, and their connections, and an exploiting agent, tasked with selecting actions and their targets. Their method partially addresses the scalability issues of DQN and demonstrates improved effectiveness for large networks.

Inspired by the notion of breaking down the PT process, Nguyen et al. introduced the DAA. DAA divides the PT problem into two steps: understanding the network structure (network topology) and selecting an exploitable service to attack a specific host. When applied to RL, instead of a single agent, DAA employs two separate agents, each handling a different step. This approach significantly reduces the action space and state space of the PT problem for each agent. Consequently, DAA is suitable for addressing challenges involving a large number of hosts and services.

We want to focus on the core principle of dividing sub-tasks logically among multiple agents. This methodology can be extended in a few different ways, such as adding even more agents or, more simply, replacing the choice of A2C agents with something more scalable.

6.2.3 The Workhorse: Double Agent + PPO (DA-PPO)

While the double agent methodology initially proposed by Nguyen et al. [10], solves the scalability problem of DQN to some extent and has a better capacity for large networks. We improve upon this method by applying proximal policy optimization (PPO) instead of A2C for both of the agents, while still retaining the concept of having separate exploration and exploitation agents. PPO was first proposed in [12], as an advanced RL algorithm known for its convergence speed, stability, and sample efficiency. It optimizes a clipped surrogate objective function to prevent performance collapse caused by a large policy update (Figure 6.4).

PPO is a popular algorithm in the field of RL that addresses some of the challenges associated with training RL agents and has many benefits. There are a variety of benefits when using PPO:

- **Stability and Robustness:** PPO is designed to ensure stability in training by limiting the size of policy updates. This helps prevent large policy changes that could lead to instability and make the learning process more robust.
- **Sample Efficiency:** PPO typically requires fewer samples to achieve good performance compared to some other RL algorithms. This is crucial in real-world applications where collecting samples can be expensive or time-consuming.
- **Policy Gradient Methods:** PPO belongs to the family of policy gradient methods, which directly optimize the policy function. This can be advantageous in

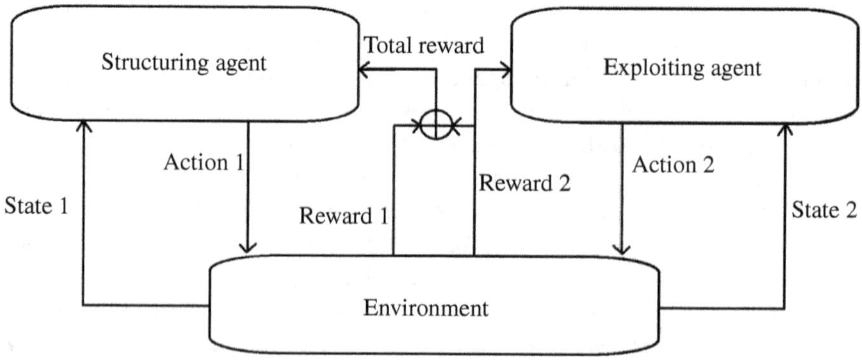

Figure 6.4 A double agent architecture has two sets of action spaces, but they both act on a single environment which has state variables separately relevant to each agent. Rewards resulting from the exploit agents actions feedback to both agents, but the structuring agents actions only directly feed rewards back to itself.

scenarios where the optimal policy is complex and hard to represent with a value function.

- **Adaptability to Different Environments:** PPO is known for its versatility and can be applied to a wide range of environments and problems. This adaptability makes it suitable for various RL tasks, from simple toy problems to complex real-world applications.
- **Simultaneous Exploration and Exploitation:** PPO incorporates a form of trust region optimization, which helps balance the trade-off between exploration and exploitation. It allows the agent to explore new actions while still exploiting the current knowledge to maximize rewards.
- **Compatibility with Deep Learning:** PPO is well-suited for deep RL, where neural networks are used to represent policies and value functions. This makes it compatible with complex and high-dimensional input spaces, such as images or raw sensor data.
- **Ease of Implementation:** PPO is relatively simple to implement compared to some other advanced RL algorithms. The algorithm's simplicity can be advantageous for researchers and practitioners looking for a reliable method without overly complex implementation challenges.
- **Parallelization:** PPO can take advantage of parallelization in the training process. This allows for efficient use of computational resources and faster training times, especially when dealing with environments that permit parallelized simulations.
- **Continual Learning:** PPO is suitable for continual learning scenarios where the agent needs to adapt to changes in the environment over time. Its incremental updates help the agent learn and adapt without forgetting previously acquired knowledge.
- **Applicability to Robotics and Control:** PPO has shown effectiveness in solving robotic control tasks, making it a valuable tool in applications where precise control and adaptation to dynamic environments are essential.

This particular architecture double agent + PPO (DA-PPO) isn't meant to be the end point of scalability, but from the size of tests alone, it is a significant leap over contemporaries. The idea is not wholly novel though, but rather a synthesis of themes seen in the various other works referenced here. The ingredients here, all borrowed from other works are:

1. Leveraging multiple agents to handle conceptually distinct subsets of actions in a **domain-informed** way.
2. Collapsing the action space to make it more manageable, in a domain-informed way.
3. Leveraging the latest advancements in scalable RL algorithms (currently PPO seems to be a dominant method).

6.3 Model Realism

6.3.1 Reward Engineering

To build a realistic simulation, the agent needs realistic motivation, and then realistic policies (and strategies and attack paths) will follow. Motivation in the context of RL is defined by the reward function and so the task becomes engineering rewards that mimic how a real attacker values actions for a given situation. This situational awareness of the attacker can be analyzed through the concept of terrain. In the context of intelligence preparation of the battlefield (IPB), "terrain" refers to the physical and geographic features of the area where military operations are planned to take place [17]. Terrain analysis is a critical component of IPB, which is a systematic process used by military intelligence to gather, evaluate, and interpret relevant information about the environment in which military forces will operate. The goal is to provide commanders with a comprehensive understanding of the battlefield to support decision-making (Figure 6.5).

Terrain analysis includes the study of various aspects of the landscape, such as:

- **Topography:** This involves examining the elevation, slopes, and contours of the land. Understanding the topography helps in identifying key vantage points, potential choke points, and areas that offer cover or concealment.
- **Vegetation:** The type and density of vegetation can impact visibility, mobility, and the effectiveness of different military systems. Dense forests, for example, may restrict movement and line of sight.
- **Urban Areas:** Analysis of urban terrain is crucial for military operations in built-up areas. Factors such as street layouts, building structures, and the presence of civilian populations are considered.
- **Hydrography:** The study of water features, including rivers, lakes, and swamps, is important for understanding mobility, potential waterborne threats, and obstacles.
- **Obstacles:** Identification of natural and man-made obstacles, such as cliffs, rivers, bridges, and roads, is essential for planning movement and defensive operations.
- **Weather and Climate:** While not strictly terrain, weather and climate conditions can significantly affect the battlefield. This includes considerations such as visibility, temperature, and precipitation.
- **Soil Composition:** The type of soil can impact mobility, especially for vehicles and infantry. Soft or muddy terrain may slow or impede movement.
- **Cultural and Human Geography:** Understanding the human aspects of the terrain, including population centers, cultural considerations, and local infrastructure, is important for shaping military plans and operations.

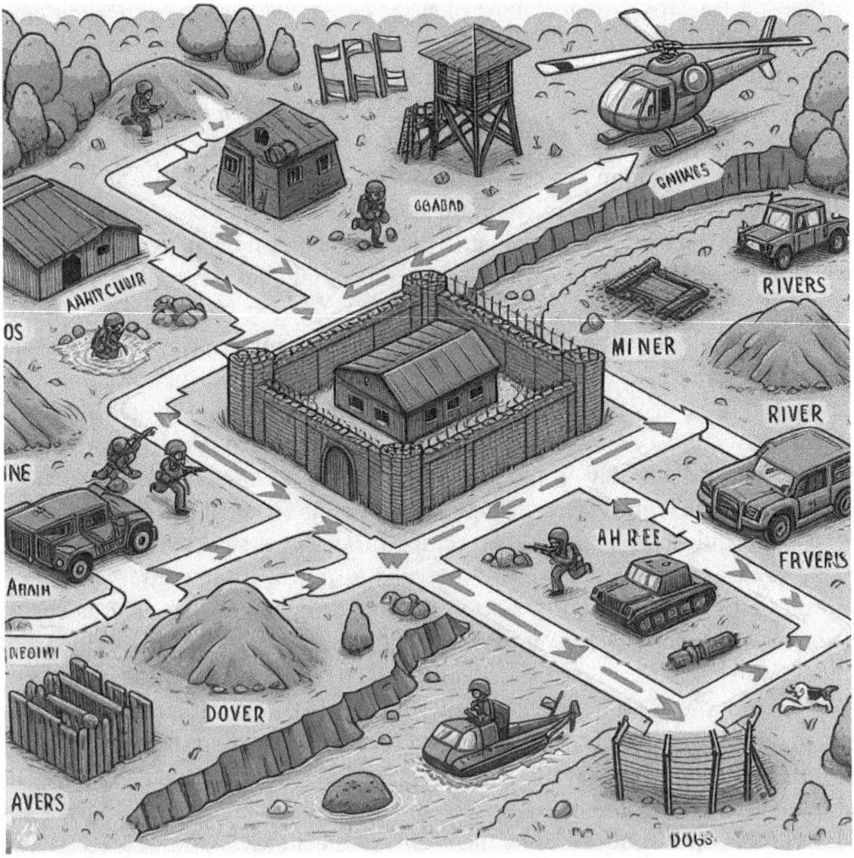

Figure 6.5 Navigating terrain in traditional and cyber warfare has many parallels. The information and options available are asymmetric between attackers and defenders. Defenses and obstacles may be obvious or implicit, and proper reconnaissance, planning and strategy can greatly effect the success of a mission.

This is a varied list, but note how there are some reoccurring themes of visibility and mobility, in fact, a lot of these considerations here have analogs in a cyber setting. Local infrastructure in a physical battle can be taken advantage of and repurposed for the attacker's use; in a cyber context, this is exactly the definition of "living off the land."

In military terminology, an "avenue of approach" refers to a route or path that an attacking force can use to reach its objective or target. This concept is particularly important in the planning and execution of military operations, where commanders and planners analyze the terrain to identify and select the most favorable avenues of approach based on factors such as cover, concealment, and terrain features. In the context of cyber operations, this is exactly the kind of attack path we want to discover using RL. In bringing in concepts of terrain and at a deeper level, how an attacker thinks about the terrain and the overall operation, our methodology comes back to the concepts introduced in LRM-RAG [3], which describes the terrain and APT layers to building the MDP. In addition to these layers, the original work on LRM-RAG describes a task-specific layer, in this context, the task is the mission, the objective is the simulation, which could be a specific type of attack campaign, or a subset of a campaign, perhaps the goal is just to conduct recon, or we want to simulate only the exfiltration piece of a larger attack campaign.

What contemporary approaches often lack by focusing only on CVSS or other "canned" metrics is a sense of context, which limits how realistic the simulation can be, but if the task of reward engineering is informed by the layered approach of LRM-RAG, this contextually aware behavior can be brought into the model. Rewards depend not only on the properties of the current node where an action is being performed but also on neighboring nodes, as an agent avoids obstacles in the terrain, and when it does make a compromise, it does so in a measured way.

The rewards in these models are designed to mirror a thoughtful consideration of the terrain, aiming to evade anticipated defenses. Instances of defensive measures encompass end point protection (EPP), anti-virus (AV), anti-malware, anti-spyware, host-based intrusion detection (HID), host-based intrusion prevention (HIP), and end point detection and response (EDR) agent technologies that can be implemented on hosts. The process involves evaluating the consequences of navigating paths that may have defenses in order to inform decision-making. In addition to these considerations, the way different actions are valued in navigating the terrain is further informed by the profile of the attacker (the APT layer) and the overall goal of the campaign (the task layer). The APT layer can be something as detailed as considering the *modus operandi* of a specific named APT, knowing which protocols, tools, and tricks they employ most frequently. Alternatively, the APT layer can describe a more generic attacker profile. Is this a "script kiddie," a loan attacker with some ready-made resources, or is this a well-funded and organized group backed by nation-state-level actors? Is this a particularly risk-averse attacker, or is it someone desperate? Both can do damage, depending on the target network, and both are worth modeling for the right end user. A model framework would describe these rewards as functions that can be scaled, tuned, or swapped when we change this attacker profile when we modify this layer. The task layer, too, can vary quite a bit; the way a particular piece of terrain is negotiated can be very different depending on the overall mission objective.

We suggest a novel method for modeling defensive terrain based on services in CVSS-MDPs. Instead of explicitly specifying defenses in the MDP states, we adopt assumptions akin to those made by human attackers. Even if an attacker cannot directly detect a defense, they can infer its presence based on the services available on a host. Common network defenses include host-based antivirus and malware detection software, inter-subnet router firewalls, and authentication log tracking. While the agent is initially unaware of the services on a host, it can adjust for expected defenses once scanning reveals the enumerated services. This has the twofold benefit of further delineating the attacker's realistic response to the information they would have available to them in an actual attack, and it helps to manage cases where explicit defense information is unavailable in scans.

We established a quantified negative reward (a penalty) structure to delineate the costs of attacker actions. The criteria of interest encompassed (1) a risk hierarchy applied to service categories and (2) the type of action executed by the agent on a host. To implement these criteria effectively, they were unified in a way that the agent could enumerate. This was accomplished by creating an array of actions and services and assigning an individual reward to each combination. The agent could then determine the appropriate negative reward for each action based on these two parameters.

In recent publications [4, 7, 13], services are categorized into four main types: authentication, data, security, and common. To generate a negative reward, a hierarchy of costs associated with attacking these services was applied. This hierarchy assigns authentication a reward of −6, data a reward of −4, and security and common a reward of −2. These rewards represent a blend of factors indicating the risk to organizations posed by these services. It's essential to recognize that the values of these negative rewards are relative, allowing for a collective scaling to represent different risk preferences for various organizations or operators. The absolute values used here are meaningless in a vacuum; tuning these values to work in the overall simulation is actually where a lot of the effort is in building these models, both so that they remain realistic and also so that they actually converge. Even so, the **relative** values shown here are still important in illustrating the design of good rewards in the relative potential risk of an area within a network based on context.

Moreover, it's crucial to distinguish between scanning and exploiting. During a scanning action, the agent incurs a negative reward with values of $n + 1$ (−1, −3, and −5, respectively). In instances where an action is taken on a host with multiple services, the agent assigns the highest cost to the reward for that action. This decision is based on the assumption that security practitioners typically adopt a best-practice approach, applying security controls based on the riskiest service identified on a host – namely, a service known to be more exposed to the network edge or associated with greater business loss if exploited. By aligning our rewards with this presumption, the agent calculates a more realistic quantitative measure of risk as it strives to converge toward an optimal attack path. It's important to note that these engineered state-action rewards for defensive terrain are additive, meaning they are layered on top of the existing CVSS-MDP rewards.

Rewards of this style can further be made contextual but scaled by proximity to the other hosts with high-negative rewards. On top of this re-scaling by proximity, the rewards can be further modified by the APT and task layers. As a simple example, for a more risk-averse attacker profile, the APT layer may dictate a global scaling of all negative rewards to higher values. In a more nuanced way, if a particular APT is known to have a protocol preference due to choice of tools or otherwise, this can also be reflected in a modification of the existing rewards already designed. In this way, the cumulative effect of these different layers in reward engineering isn't simply to add different rewards for different situations, though we will do that as well.

6.3.2 Human Inputs vs. Model Inputs

A large benefit of the methodology proposed here is that the simulation occurs entirely offline. Once the initial scans are complete, there is no impact on the network for which an attack is simulated. Not only does this allow pen testing to occur with minimal disruption, but it also allows for testing new network configurations and defensive strategies virtually, without the same investment as making the changes in reality. To achieve this, we want to represent and build the environment for simulation in such a way that it contains all the useful information from scans so that it is realistic, but also have it organized well enough that a synthetic network can be created whole cloth if desired.

In the current state of this methodology, the primary scan information comes from Network Mapper (Nmap). Nmap is a powerful open-source network scanning tool that is widely used for network discovery and security auditing. An Nmap scan can provide detailed information about hosts on a network. The specific information contained in an Nmap scan depends on the type of scan and the options used, but here are some common elements that an Nmap scan may reveal:

- **Open Ports:** Nmap scans identify open ports on a target system. Ports are entry points for network services and knowing which ports are open can help identify potential vulnerabilities or services running on the target.
- **Service Versions:** Nmap can often determine the version of a service running on an open port. This information is valuable for understanding the software and its potential vulnerabilities. For example, it might reveal the version of an HTTP server or an SSH daemon.
- **Operating System Detection:** Nmap can attempt to determine the operating system of the target system based on subtle differences in the way different operating systems respond to network probes.

- **Host Discovery:** Nmap can be used to discover hosts on a network. It can identify live hosts and determine which hosts are currently available.

While at first glance, this may seem somewhat limited, this actually provides a good chunk of the information required to plan an attack; understanding what hosts are present and what operating systems and services are on each can tell an attacker a lot about how they would jump between hosts and exploit each. This information is far from complete; however, in particular, there is no complete explicit information on the topology of the network. A lot about topology can be inferred. Nmap can provide some level of explicit topology information about a network, but it primarily excels at discovering hosts and open ports rather than providing a detailed network topology map. Here are some aspects of topology-related information that Nmap can reveal:

- **Host Discovery:** Nmap can identify live hosts on a network. It uses various techniques, such as ping probes or ARP requests, to determine which hosts are active.
- **Open Ports:** Nmap provides information about open ports on a host. While this doesn't directly give you network topology, it does reveal the services running on different hosts, which can be a starting point for understanding how hosts are interconnected.
- **OS Fingerprinting:** Nmap has the capability to perform OS fingerprinting, attempting to identify the operating system of each host. While this doesn't explicitly give you a topology, knowing the different operating systems in use can provide insights into the diversity of devices on the network.
- **Traceroute:** Nmap includes a traceroute feature (-traceroute option) that can be used to identify the path that packets take from the scanning machine to a target host. This can give you a sense of the routing topology between the scanning machine and the target host.
- **Topology Mapping Scripts:** Nmap includes scripting capabilities that allow users to create custom scripts for additional functionality. There are scripts designed to provide more explicit topology-related information, such as snmp-interfaces or snmp-netstat, which can query SNMP-enabled devices for network interface information.

While Nmap can provide valuable information, it's important to note that mapping an entire network's topology may require additional tools or a combination of different techniques. Tools like OpenVAS, Nessus, or specialized network mapping tools may be more suitable for comprehensive network topology discovery. Regardless, Nmap is considered the foundational tool for collecting information for this type of simulation and will inform the data structure of the environment.

To illustrate how this information is gathered into a single file, it is useful to walk through the process of generating a synthetic network. To generate realistic network topologies for testing, it is imperative to ensure that networks exhibit realism in terms of size, scope, composition, connectivity, and security posture. The specified requirements are as follows:

- Networks must realistically reflect their size and scope.
- Networks must be realistic in their composition.
- Networks must exhibit realistic connectivity.
- Networks must embody a realistic security posture.

To meet these requirements, the following steps were executed during the initial network generation and configuration: A set of variables was established to govern the network's size and scope, encompassing the total (approximate) number of IPs, minimum and maximum IPs per subnet, and total number of subnets. These variables, assigned static values, were utilized by the network generation script to construct a network that closely aligns with the desired size and scope. Randomization is a crucial factor to ensure the uniqueness of each generated network. Another set of variables was defined to configure the network according to the desired specifications, including the maximum number of open ports and common platform enumeration (CPE) types assigned to an IP address. Randomization is again vital to ensure uniqueness.

A dataframe was constructed to store continuous 24-hour data collected from the Shodan API [18]. This dataframe was employed to generate a reference dataframe, linking CPEs by service and technology to each IP address. Using a group by function, the results were aggregated by port, service, and technology, forming the reference dataframe containing port numbers, lists of service and technology combinations, and associated probabilities. This reference dataframe was saved separately for persistent use.

Another dataframe was created to include ports and their probabilities, determined by the probability score of each port based on the open-frequency score in the "Nmap services" publication.

To integrate these dataframes and variables into a cohesive network, a series of algorithms were devised, implementing the network's configuration and ensuring the desired connectivity. The algorithms employed include:

- Assignment of IP addresses to each subnet based on specified size and scope settings.
- Determination of the number of ports assigned to each IP address, with randomization ensuring variability.
- Assignment of ports to IP addresses based on randomly generated probability scores.
- Utilization of the CPE dataframe to assign CPEs to relevant open ports of each IP address, with consideration for probability scores. Enrichment of the CPE

dataframe with a new column associating known security products with corresponding IP addresses.

At this stage of the network generation workflow, the network comprises a defined number of subnets, multiple IP addresses assigned to each subnet, open ports assigned to each IP address, and CPEs assigned to the necessary open ports. To fulfill the desired connectivity of the network, an algorithm was devised to assess the viability of service connections between different subnets, implementing a firewall allowing rules based on desired connectivity settings.

Following the testing of viable network paths, it was observed that the abundance of path options between entry and exit points resulted in scenarios that were unrealistic and oversimplified. To address this, a manual review of the initial testing networks led to the removal of certain subnet connections, enhancing realism. As a next-step enhancement, a new algorithm will be developed to limit the number of paths between entry and exit points, randomly selecting a subset based on a reference dataframe containing common firewall rules and their probability scores.

To introduce vulnerabilities into the network, CVE exploits are assigned to each CPE. The CVE-Search database is queried for this information, with each CPE queried against the database to capture the corresponding CVE exploits. The resulting CVE exploits, along with their associated CVE ID, CVSS score, and CVSS vector, are attributed to the CPE record in a new dataframe, which is then saved for persistent use. Note that a newly developed Python library, PyCVESearch, will replace the current method of maintaining and querying a local CVE-Search database, allowing for a more efficient query mechanism. The CVE-Search database also provides the option to query for CWE information. This information is used to determine the type of hardware, software, or architectural weakness associated with a given CVE, contributing to a more realistic representation of exploitable vulnerabilities and their shared inherent weaknesses in the network.

Once all elements of the network are created, the output is consolidated into a singular YAML file. This file encompasses all necessary information for creating testing scenarios, including the number of subnets, IP addresses per subnet, open ports per IP address, associated services and technologies per open port, CPEs assigned to open ports, CVE exploits assigned to CPEs, CVSS scores, CVSS vectors, firewall allow rules between subnets, and designated security products.

While no explicit attack graph is computed, a topology from the information Nmap provides, as described above, and this basic topology is represented as a single matrix in the YAML file that will inform the agent when a host is discoverable and what hops are possible. Again, it is important to differentiate between the information available to the model (in this case, we explicitly mean all the information that's been brought into the YAML file), and the information that is known to the agent at any one time (through the results of its scanning actions). This YAML file describing a wholly synthetic network contains exactly the same information as YAML file that is generated from real scan data, and much the same procedure is followed, except where explicit information form scans is available, it is used instead.

6.4 Examples of Applications

The flow of this book up to this point has been from more general to more specific in how we simulate attack campaigns with RL. Starting with the basic backgrounds of pen-testing and RL, respectively, and then slowly building the framework to a realistic, scalable modeling approach. In a layered approach, inspired by LRM-RAG, the task layer doesn't just add new actions and rewards, but it modifies the other layers as well. The next chapter will go into more depth into how different types of modeling tasks are considered in our framework, but for now, we give a brief overview.

6.4.1 SDR

Reconnaissance, commonly referred to as "recon," plays a crucial role in cyber operations and is often the first phase in the cyberattack life cycle. The primary objective of reconnaissance is to gather information about the target system, network, or organization, enabling the attacker to understand the target's environment, vulnerabilities, and potential points of entry. Emphasis on establishing situational domain awareness (SDA) is vital during the reconnaissance phase. Again, **context** is important. Business leaders and even CISOs may be less concerned with simulating recon, but since it is usually a required first part of a real attack campaign, this is potentially the earliest an attacker can be stopped in their tracks before they have caused harm. Mostly, early prevention is focused on the attack surface, but no attack surface is perfectly defended; after an adversary enters a network, they will most likely be gathering information from the inside before they proceed further. An important feature of this information gathering is that it is **asymmetric**, the attacker wants to observe without being observed. Again in analogy to military terminology, these paths of asymmetric information are quite similar to surveillance detection routes (SDRs). SDRs are pre-planned routes designed to help individuals or security teams identify potential surveillance or threats. These routes are commonly used in security and counter-surveillance practices to enhance situational awareness and detect signs of unwanted or suspicious attention. The primary goal of an SDR is to allow individuals to observe and identify surveillance activities, providing them with the opportunity to respond appropriately.

Key features of SDRs include:

- **Variability:** SDRs should have multiple possible routes, schedules, and destinations. This variability makes it more challenging for potential surveillants to predict the subject's movements, increasing the effectiveness of the detection process.
- **Frequency:** SDRs are typically conducted regularly to establish a baseline of normal activity. Regularity allows individuals or security personnel to better recognize deviations or patterns that may indicate surveillance.
- **Inconspicuous Observation Points:** SDRs often include stops or activities in areas where individuals can discreetly observe their surroundings without drawing attention. This may involve choosing locations that offer good visibility, such as cafes, parks, or other public spaces.
- **Awareness Training:** Individuals following an SDR are trained to be observant and aware of their surroundings. This includes recognizing behavioral cues, identifying unusual activities, and noting potential indicators of surveillance.
- **Unpredictable Elements:** SDRs may incorporate unpredictable elements, such as sudden changes in direction, stops, or modes of transportation. These elements help disrupt any potential surveillance pattern and reveal the presence of an observer.
- **Communication Protocols:** Individuals or security teams conducting SDRs often have established communication protocols. This may include a system for reporting and responding to potential surveillance incidents in real-time.
- **Documentation:** Keeping detailed records of SDRs, including routes taken, observations made, and any identified anomalies, is crucial for analysis and future planning. Documentation aids in recognizing patterns over time.
- **Collaboration with Security Teams:** In organizational settings, SDRs may be part of a broader security strategy that involves collaboration with dedicated security teams. These teams can provide additional support and expertise in surveillance detection.

SDRs are commonly employed by individuals who may be at higher risk of surveillance, such as diplomats, corporate executives, or high-profile personalities. Additionally, security teams for critical infrastructure, government agencies, or private organizations may incorporate SDRs as part of their overall security measures. In the case of a cyber attacker, much of the same considerations are made. The goal of the SDR simulation is to find a path through the network where an attacker can gather as much information as it can (run scans on as many hosts as possible) while evading detection. It should be obvious that it requires a

Figure 6.6 A surveillance detection route is a method for detecting potential surveillance activities or put another way, a method for surveillance of surveillance. This is done by exploiting an asymmetry in information. The hooded figure with the spy glass is trying to gain information about where he may be detected. The dark alley (a possible attack path) is dotted with street lights, a patrol car, and a watch dog, various forms of defense the figure would have to have a plan to avoid if he were to navigate the alley.

balancing act of positive and negative rewards that will need to be engineered to make these paths realistic (Figures 6.6 and 6.7).

6.4.2 Crown Jewels Analysis

After an initial foothold and recon, the "main" part of a mission can be undertaken, and this usually entails some variation on gaining control of a particular host: a classic "capture the flag" scenario. Perhaps, the information on this host will be exfiltrated, or maybe it will be encrypted as part of a ransomware

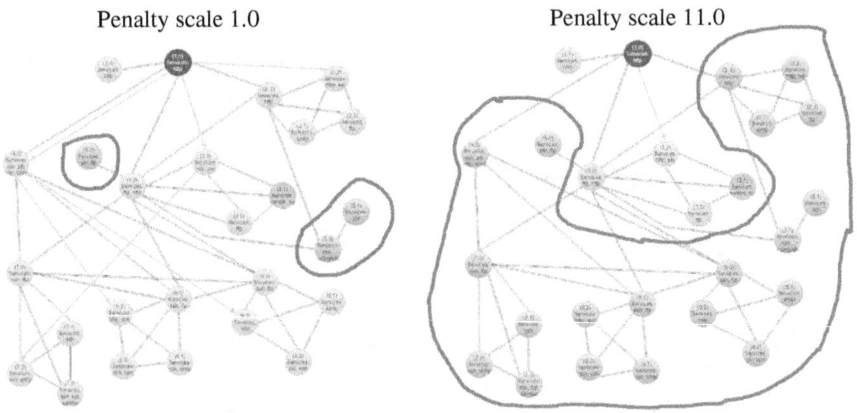

Figure 6.7 As the penalty scale is increased, an agent acts in a more risk-averse manner. In the case of a reconnaissance mission, the agent will explore less of the network.

attack. There are many possibilities, but if there is one or a few particular hows which are understood by an organization to be very valuable, these are often referred to as crown jewels. Crown Jewel Analysis is a security risk assessment methodology that focuses on identifying and protecting an organization's most critical assets or "crown jewels." The term "crown jewels" refers to the most valuable, sensitive, and crucial assets within an organization that, if compromised, could have a severe impact on its operations, reputation, and overall well-being. A simple "capture the flag" simulation can add a lot to crown jewel analysis as it (again) provides a lot of contextual information. Even if all the atomic stepping stones leading up to the crown jewels seem secure, a path may exist between them, exploiting seemingly low-risk vulnerabilities (Figures 6.8).

6.4.3 Discovering Exfiltration Paths with RL

Once an attacker has gained control of the target host(s), the final step is often exfiltration. Exfiltration, in the context of cybersecurity, refers to the unauthorized and often surreptitious extraction or transfer of data from a computer system or network by an attacker. The term is commonly used to describe the process by which sensitive or valuable information is illicitly taken from an organization's internal systems and transmitted to an external location controlled by the attacker. Exfiltration is a significant concern in cybersecurity as it can lead to data breaches, intellectual property theft, and compromise of sensitive information (Figures 6.9 and 6.10).

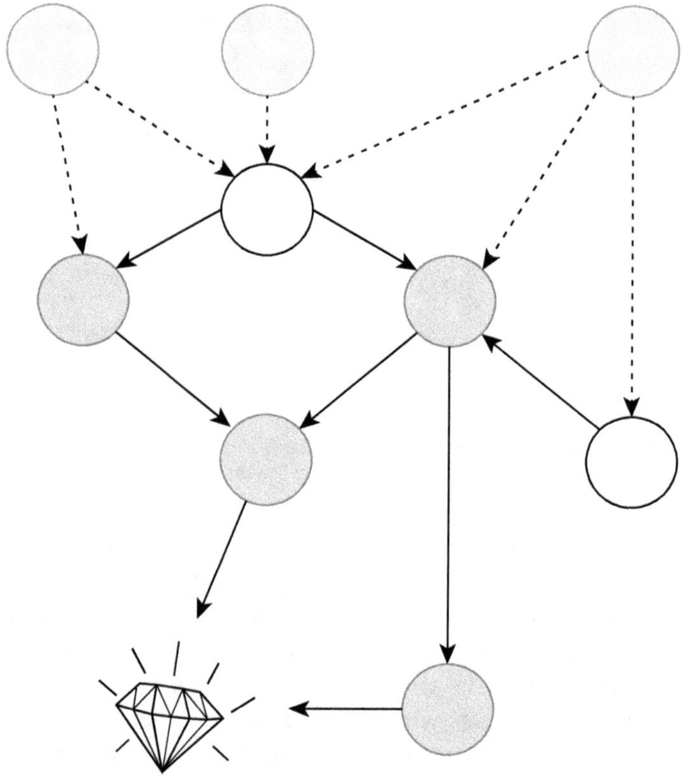

Figure 6.8 Diagram of network nodes with a few hops of a high-value target (crown jewels) blue nodes are two hops or less from the target.

On the surface, this seems like it would just be the capture-the-flag path in reverse, but the requirement of taking the payload with you to the exit and doing so undetected complicates things greatly.

Data Transfer: Exfiltration involves the unauthorized transfer of data from the victim's network or system to a location outside the organization's control. This transfer may occur through various means, including network communication channels, file transfers, or other covert methods.

Methods of Exfiltration:

- **Network-based Exfiltration:** Attackers may use network protocols to transmit stolen data outside the organization, often disguising the communication to evade detection.
- **Covert Channels:** Exfiltration can occur through hidden or less monitored communication channels, such as DNS requests, covert channels in network protocols, or seemingly innocuous traffic.

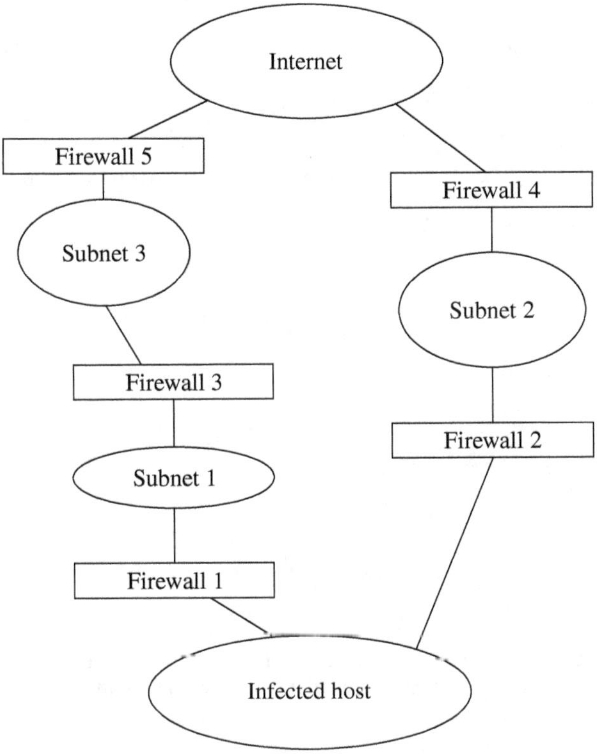

Figure 6.9 The shortest path to exfiltrate data from an infected host to the internet is often safer, as the path will cross fewer defenses, such as the path to the right above, which only crosses fewer firewalls than the path to the left.

- **Removable Media:** Attackers might physically transfer data using removable storage devices like USB drives or external hard disks.
- **Steganography:** Exfiltration may involve hiding data within seemingly harmless files or communication to avoid detection.

Targets of exfiltration include sensitive and valuable information such as personally identifiable information (PII), financial data, intellectual property, trade secrets, or classified information. Alternatively, exfiltration targets could include various sorts of credentials: Stolen usernames and passwords may be exfiltrated to gain unauthorized access to systems or services.

1. Detection Challenges: Exfiltration attempts are often designed to be stealthy and evade detection by security mechanisms. Advanced attackers may use encryption, obfuscation, or other evasion techniques to avoid detection during data transfer.

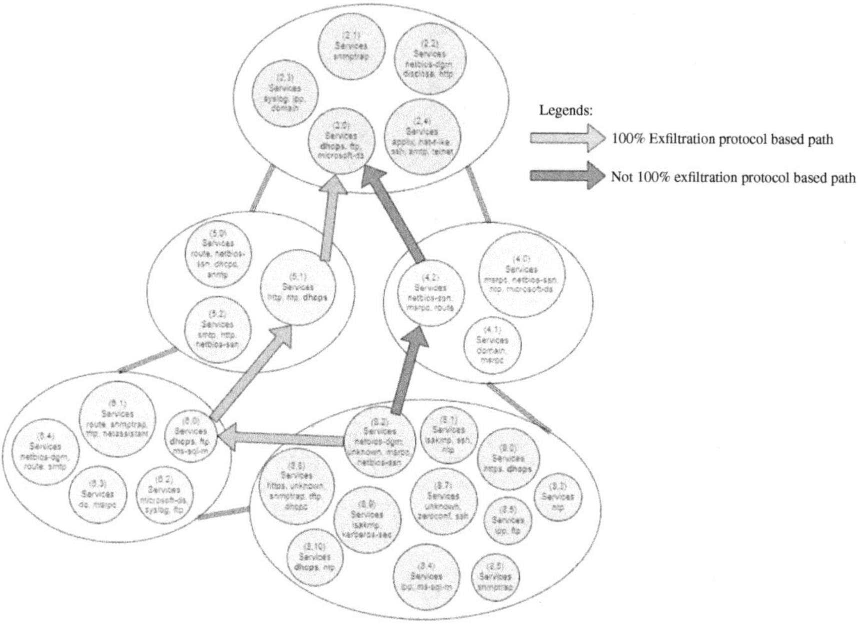

Figure 6.10 An attacker that sticks to a particular protocol preferentially may see longer exfiltration paths, but they may also avoid detection by blending in with benign or otherwise unmonitored traffic over the network.

2. Prevention and Mitigation: Organizations employ various cybersecurity measures to prevent and mitigate exfiltration, including intrusion detection and prevention systems (IDPS), data loss prevention (DLP) solutions, encryption, network monitoring, and user training on security best practices.
3. Insider Threats: Exfiltration may occur due to insider threats, where individuals within an organization intentionally or unintentionally facilitate the unauthorized transfer of data.
4. Post-Compromise Activities: Exfiltration is often part of the post-compromise phase of a cyberattack, where attackers seek to capitalize on the initial access gained to achieve their objectives, such as stealing valuable information or disrupting operations.

Detecting and preventing exfiltration requires a combination of technical controls, user awareness, and proactive security measures. Organizations need to continuously assess and improve their cybersecurity posture to defend against evolving exfiltration techniques employed by cyber adversaries. Moving a network's payload stealthily will require its own balancing act of rewards and penalties.

6.4.4 C2

Command and control (C2), in the context of cybersecurity, refers to the communication and coordination infrastructure used by attackers to manage and control compromised systems or networks. The term is commonly associated with malicious actors, such as hackers or cyber criminals, who employ C2 systems to remotely operate and manipulate compromised devices for various malicious purposes. Understanding C2 is crucial for cybersecurity professionals to detect, prevent, and respond to cyber threats effectively.

Key aspects of C2 in cybersecurity encompass various elements. The primary objective of a C2 system is to provide attackers with remote control over compromised devices or networks, enabling the execution of commands, data extraction, and other malicious activities without the need for direct physical access.

The components of a C2 infrastructure include the command server, acting as the central communication hub, where attackers dispatch instructions to compromised systems. Controlled agents, representing the compromised devices or systems, connect to the command server to receive directives and report back on their status.

Communication channels within C2 systems often employ covert or disguised methods to evade detection. This may involve using legitimate protocols, encrypting communications, or employing steganography techniques. Attackers may leverage common network protocols like HTTP, DNS, or custom protocols to camouflage C2 traffic amidst normal network activity.

C2 facilitates functions and activities such as command execution, where attackers issue directives to download and execute malicious payloads, exfiltrate data, or perform reconnaissance. C2 servers also play a crucial role in enabling updates to malware, changes in attack tactics, or adaptation to evolving security defenses, allowing attackers to maintain control over compromised infrastructure.

Detection of C2 activities poses challenges due to evasion techniques employed by attackers, including encrypted communication, mimicking legitimate traffic patterns, and periodic changes to C2 infrastructure. Another example is polymorphic malware, with the ability to alter its code structure to evade signature-based detection, further complicates identification by security solutions.

Mitigation and prevention strategies involve continuous network monitoring and anomaly detection to identify unusual communication patterns associated with C2 activity. Behavioral analysis by security solutions focusing on the behavior of applications and systems aids in detecting suspicious activities are indicative of C2 communication. Proactive measures include blocking C2 traffic, disrupting communication channels, and isolating compromised systems to prevent further malicious actions.

Understanding the tactics and techniques used in C2 operations is essential for developing effective cybersecurity strategies. It enables organizations to detect and respond to C2 activities promptly, minimizing the impact of cyber threats on their systems and data. In a simulation of C2, new actions are introduced for establishing a connection to that outside controller and uploading to it, and once this upload action has been done for each host in a list of sensitive hosts the simulation is considered successful.

6.4.5 Ransomware

Ransomware is a type of malicious software (malware) designed to block access to a computer system or files until a sum of money, or ransom is paid by the victim. It is a form of cyber extortion where attackers encrypt the victim's data and demand payment in exchange for providing the decryption key or restoring access to the affected system. Ransomware attacks can target individuals, businesses, and even government entities, and they have become a significant and widespread cybersecurity threat (Figure 6.11).

Ransomware typically employs data encryption, encrypting a victim's files or entire hard drive, making the data inaccessible without the decryption key. The use of advanced encryption standards, such as AES, by attackers, adds complexity to decryption efforts. Ransom demands are often made in cryptocurrency, like Bitcoin, to enhance transaction traceability difficulty, and victims receive ransom notes detailing the payment process and instructions for obtaining the decryption key. Delivery methods include phishing emails with malicious attachments or links and visits to compromised or malicious websites, leading to ransomware download and execution.

Ransomware targets both individuals for personal gain and businesses for financial extortion, with critical infrastructure, healthcare organizations, and government entities occasionally falling victim, causing widespread disruption. The evolution of ransomware includes the use of Ransomware-as-a-Service platforms by some attackers, enabling less technically skilled individuals to launch attacks, and the tactic of double extortion, where attackers threaten to publicly release sensitive data unless the ransom is paid.

Prevention and mitigation strategies involve regular data backups, allowing organizations to restore systems in case of an attack, the use of robust antivirus and anti-malware solutions to detect and prevent infections, and user education to recognize phishing emails and practice safe online behavior. Ransomware attacks are illegal, and paying the ransom offers no guarantee that attackers will fulfill their promises, leading to ethical dilemmas for organizations and individuals in deciding whether to resist extortion attempts or comply with ransom demands.

Figure 6.11 A diagram of a typical incident of a ransomware attack. An otherwise healthy machine (or network of machines) acquires malware in some way which then phones home to an attacker who in turn encrypts high-value assets which will remain encrypted until the attacker receives payment from their victims.

Ransomware attacks have had significant financial and operational impacts on individuals and organizations worldwide. As a result, cybersecurity measures and awareness are crucial to defending against and mitigating the risks associated with ransomware threats.

To simulate ransomware obviously requires a new **encrypt** action, but also, in simulating ransomware, we attempt to model a common defense against it: honeypots.

A honeypot in cybersecurity is a security mechanism set up to attract and detect attackers or unauthorized users. It is a deceptive system designed to mimic real systems and networks, luring potential attackers into interacting with it. The primary purpose of a honeypot is to gather information about the tactics, techniques, and procedures (TTPs) used by attackers and to divert and study malicious activities away from production systems. Honeypots are valuable tools for cybersecurity professionals to enhance threat intelligence and understand the current threat landscape.

Honeypots exhibit key characteristics designed to deceive and entice potential attackers. These simulated systems, networks, and services mimic real environments, with some employing fabricated data to enhance their allure. Honeypots come in various types, including research honeypots utilized by cybersecurity researchers for studying new attack techniques, production honeypots deployed within an organization's environment to detect and deflect attacks, low-interaction honeypots simulating a limited set of services, and high-interaction honeypots emulating complete operating systems and services for a more realistic environment.

The goals of honeypots encompass information gathering on attackers' methods, tools, and intentions, providing early warnings to security teams about potential threats or ongoing attacks, and diverting attackers away from actual production systems to minimize damage. Honeypots can be deployed as network honeypots, simulating entire networks to attract attackers scanning for vulnerabilities, or as host-based honeypots, emulating individual systems or services to attract specific types of attacks.

Challenges associated with honeypots include the potential generation of false positives when legitimate users or automated scanners interact with them, necessitating regular updates and monitoring for ongoing effectiveness. Legal and ethical considerations are paramount, requiring authorization to ensure compliance with standards and adherence to privacy regulations in handling and storing data collected by honeypots.

Honeypots find applications in diverse use cases, contributing valuable information to threat intelligence databases for staying informed about emerging threats and aiding incident response by providing insights into attacker behavior and tactics.

6.5 Realism and Scale

The varied use cases in the previous section should all hint at how reward engineering can be a complicated process. Understanding not just the motivations of an attacker but also the nuances and trade-offs in those motivations. At a broad level, this is often a trade-off between speed and stealth, but there are often numerous trade-offs to consider. Much of the "work" in getting different functionalities to work in RL is in reward engineering, both in the drawn-out iterative conversations between data scientists and cyber subject matter experts. These conversations ensure the **concepts** are motivated correctly so that reward engineering is properly constructed. In recent works, [4, 7, 13], the values for these various positive and negative rewards are tuned quite manually. Specifically, the primary objective is finding values for the reward that are scaled proportionally so that training converges on an actual useable policy. To make this even **more** complicated, some of these rewards may require end-user tuning while maintaining auto-balancing. Based on this, it is a better defensive strategy to anticipate different kinds of attackers that will make different choices in these trade-offs. Some threat actors may be able to do months-long operations and work slowly; others may have no opportunity other than a "smash and grab" attempt. A good defensive posture will support readiness in all cases. In addition to all the variability in the tasks above and the inherited trade-offs involved, these simulated scenarios are not useful if they produce a single attack path but, rather, a distribution of behavior. To meet these challenges and to continue to scale this kind of model, not just to run on larger networks but to be extensible to many different use cases quickly, we turn to two promising and interrelated techniques: multi-task learning and multi-objective learning.

6.5.1 Multi-task Learning

Multi-task learning (MTL) in the context of neural networks refers to a machine learning paradigm where a single model is trained to perform multiple tasks simultaneously. In traditional machine learning approaches, different models might be trained for different tasks, but in multi-task learning, a single neural network is designed to handle multiple related tasks.

The main idea behind multi-task learning is that the shared representation learned by the model can benefit the performance of each individual task. By jointly learning multiple tasks, the model can exploit commonalities and relationships between the tasks, which can lead to improved generalization and better overall performance.

In a neural network architecture for multi-task learning, there are typically shared layers that are common to all tasks, and then there are task-specific layers that are unique to each task. The shared layers capture the shared features and representations, while the task-specific layers focus on learning the details specific to each task. During training, the loss function is a combination of the losses from all tasks, and the back propagation algorithm adjusts the parameters of the model to minimize this combined loss.

Benefits of multi-task learning include:

- **Improved Generalization:** The shared representation helps the model generalize better to new, unseen data for each task.
- **Data Efficiency:** Training a single model for multiple tasks can be more data-efficient than training separate models for each task, especially when the amount of data for individual tasks is limited.

- **Regularization:** MTL can act as a form of regularization, preventing overfitting by encouraging the model to focus on the most informative features.
- **Transfer Learning:** Knowledge learned from one task can be transferred to another task, especially when tasks are related.

However, designing an effective multi-task learning architecture can be challenging, as there needs to be a balance between shared and task-specific layers to ensure optimal performance for each task. The choice of tasks and how they are related also plays a crucial role in the success of multi-task learning. When all layers are shared and balancing all occur in the loss function, then the loss is often represented as a linear combination of the losses for the sub tasks and the coefficients in this sum are parameters that the model can learn as well. In this fashion, much of the tuning of these tasks may itself be data driven-in the future.

MTL in the context of RL involves training a single agent to perform multiple related tasks simultaneously. This approach is motivated by the idea that learning multiple tasks concurrently can lead to better generalization and more efficient use of the agent's experience.

In the realm of multi-task RL, a common strategy involves employing a shared policy and/or value function across various tasks. This entails utilizing a neural network to represent the policy or value function, with specific layers shared among tasks to capture common features and representations. This shared network component is dedicated to learning general features applicable to all tasks. Complementing the shared layers are task-specific modules, integrated into the neural network architecture to capture the intricacies of each individual task. During training, the agent undergoes transitions from all tasks, with the learning algorithm adjusting the parameters of both shared and task-specific layers to maximize cumulative rewards. The learning signal from each task contributes to a joint objective function, guiding the agent in optimization. Multi-task RL also facilitates transfer learning, leveraging the shared representation as a transferable knowledge base for quicker adaptation to new tasks. Task selection is pivotal, demanding a delicate balance between relatedness to exploit shared features and diversity to provide meaningful challenges. Some approaches incorporate adaptive learning mechanisms to dynamically adjust the importance of each task during training, allowing the agent to focus on challenging or sparsely rewarded tasks. Multi-task RL training methods may involve sequential or parallel approaches, interleaving experiences or updating the model simultaneously across different tasks. Applying multi-task learning to RL is an active area of research, and the effectiveness of the approach can depend on the specifics of the tasks, the shared representation, and the learning algorithm used. It offers the potential to improve sample efficiency, enhance generalization, and enable agents to learn a diverse set of skills.

6.5.2 Multi-Objective Learning

A very closely related topic to multi-task learning is multi-objective learning. Multi-objective learning (MOL), also known as multi-objective optimization or multi-objective reinforcement learning (MORL), is a paradigm in machine learning where a model is trained to simultaneously optimize multiple objectives or criteria. In contrast to single-objective optimization, where the goal is to find a single solution that optimally balances all considerations, multi-objective learning deals with the challenge of finding a set of solutions that represent a trade-off between conflicting objectives.

In the context of neural networks and optimization problems, multi-objective learning involves training a model to optimize multiple objective functions concurrently. Each objective function corresponds to a different goal or criterion, and the aim is to find a set of solutions that collectively represent a compromise or Pareto front - no solution is better in all objectives, but there are solutions that are better in some objectives than others.

MOL involves several key aspects, beginning with the definition of multiple objective functions, each representing different aspects or goals of a given problem. These objectives may be conflicting, where improvements in one objective could potentially compromise another. The solutions that represent optimal trade-offs between these objectives form the Pareto front, and a solution is deemed Pareto optimal if no other solution surpasses it in all objectives. Pareto dominance is a criterion in multi-objective optimization, stating that one solution dominates another if it is at least as good in all objectives and strictly better in at least one. MOL facilitates trade-off analysis, allowing decision-makers to choose solutions aligning with their preferences. Applied across various fields, such as optimization problems, evolutionary algorithms, RL, and decision-making systems, multi-objective learning proves invaluable in scenarios with multiple, potentially conflicting, criteria to consider. Evolutionary algorithms, like genetic algorithms, are frequently employed in multi-objective optimization, maintaining solution populations and evolving them over generations to explore the Pareto front. In the context of RL, MOL, known as MORL, trains agents to optimize multiple objectives simultaneously, particularly relevant in scenarios requiring agents to balance between different goals.

MOL is valuable in situations where there is no single, universally optimal solution, and decision-makers need to navigate trade-offs between competing objectives. It provides a framework for exploring the diverse set of solutions that exist in the trade-off space. While MOL and multi-task learning (MTL) both involve training models to handle multiple objectives or tasks, they have distinct focuses and are applied in different contexts. Here are the key differences and the relationship between multi-objective learning and multi-task learning:In the realm of machine learning, the distinction between MOL and MTL is significant. MOL involves optimizing multiple objective functions that may represent various criteria or goals, often in conflict, aiming to find solutions that strike a trade-off between these objectives. The focus here is on the Pareto front, a set of solutions representing optimal trade-offs between conflicting objectives. In contrast, MTL concentrates on training a model to perform multiple tasks simultaneously, assuming underlying commonalities between tasks and emphasizing shared representations. While MOL is associated with optimization problems and employs techniques like Pareto optimization and evolutionary algorithms, MTL is commonly applied in supervised learning scenarios, aiming for joint learning of multiple tasks to enhance generalization. MOL finds application in contexts

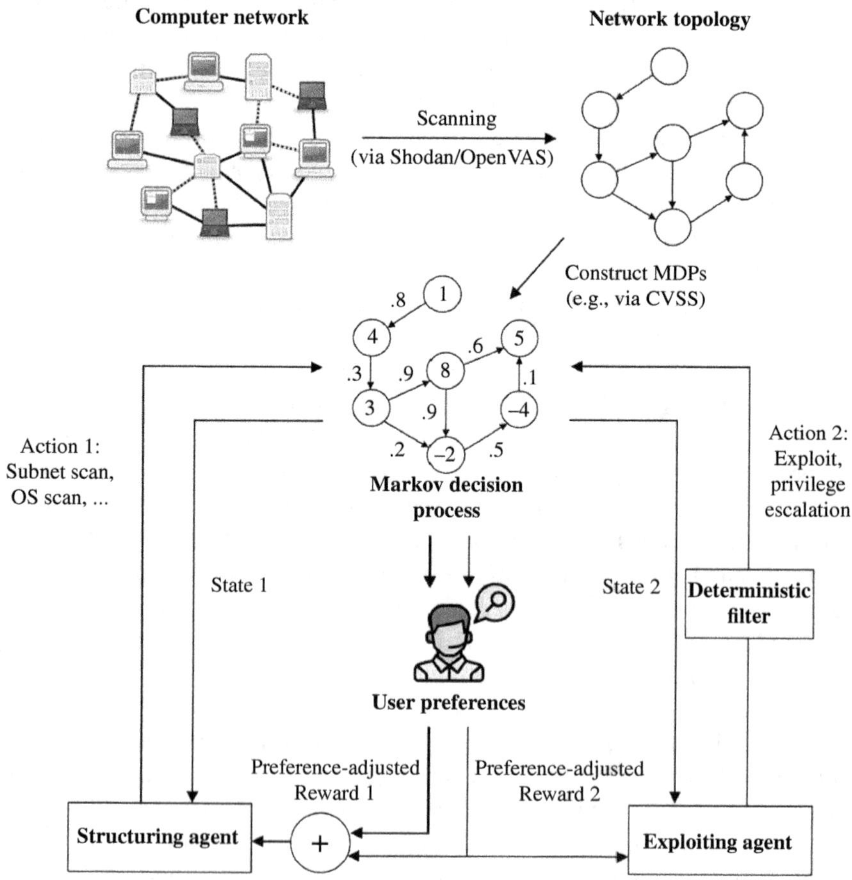

Figure 6.12 Overall view of Mult-Objective RL with preference adjusted rewards for structuring and exploiting agents. Source: Adapted from Huang et al. [8].

with conflicting objectives, such as engineering design or finance, while MTL is employed in scenarios with multiple related tasks, like natural language processing or computer vision. The learning approaches differ, with MOL focusing on diverse solutions for decision-makers to choose based on preferences and trade-offs, and MTL emphasizing shared representations for improved generalization and potential transfer learning benefits [8] (Figure 6.12).

In summary, while both MOL and MTL involve learning from multiple sources of information, MOL is more concerned with finding optimal trade-offs between conflicting objectives, whereas MTL is focused on learning shared representations to improve performance on multiple related tasks. The relationship between the two lies in the broader domain of learning from multiple sources, but the specific objectives and methodologies differ.

Applying multi-task learning and multi-objective learning to RL and penetration testing in particular is an active area of research. Meshing these techniques with the framework discussed in the rest of this book is a task not yet completed, but it is a promising path forward.

References

1 Sujita Chaudhary, Austin O'Brien, and Shengjie Xu. Automated post-breach penetration testing through reinforcement learning. In *2020 IEEE Conference on Communications and Network Security (CNS)*, pages 1–2. IEEE, 2020.

2 Ankur Chowdary, Dijiang Huang, Jayasurya Sevalur Mahendran, Daniel Romo, Yuli Deng, and Abdulhakim Sabur. Autonomous security analysis and penetration testing. In *2020 16th International Conference on Mobility, Sensing and Networking (MSN)*, pages 508–515. IEEE, 2020.

3 Tyler Cody. A layered reference model for penetration testing with reinforcement learning and attack graphs. In *2022 IEEE 29th Annual Software Technology Conference (STC)*, pages 41–50. IEEE, 2022.

4 Tyler Cody, Abdul Rahman, Christopher Redino, Lanxiao Huang, Ryan Clark, Akshay Kakkar, Deepak Kushwaha, Paul Park, Peter Beling, and Edward Bowen. Discovering exfiltration paths using reinforcement learning with attack graphs. In *2022 IEEE Conference on Dependable and Secure Computing (DSC)*, pages 1–8, 2022.

5 Mohamed C Ghanem and Thomas M Chen. Reinforcement learning for intelligent penetration testing. In *2018 Second World Conference on Smart Trends in Systems, Security and Sustainability (WorldS4)*, pages 185–192. IEEE, 2018.

6 Mohamed C Ghanem and Thomas M Chen. Reinforcement learning for efficient network penetration testing. *Information*, 11(1):6, 2020.

7 Lanxiao Huang, Tyler Cody, Abdul Rahman, Christopher Redino, Ryan Clark, Akshay Kakkar, Deepak Kushwaha, Paul Park, Peter Beling, and Edward Bowen. Exposing surveillance detection routes via reinforcement learning, attack graphs, and cyber terrain. In *2022 21st IEEE International Conference on Machine Learning and Applications (ICMLA)*, pages 1–8, 2022.

8 Lanxiao Huang, Tyler Cody, Christopher Redino, Abdul Rahman, Cheng Wang, Ryan Clark, Edward Bowen, Peter Beling, and Ming Jin. Automated preference-based penetration testing with multi-objective reinforcement learning. *unpublished*, 2024.

9 Qianyu Li, Min Zhang, Yi Shen, Ruipeng Wang, Miao Hu, Yang Li, and Hao Hao. A hierarchical deep reinforcement learning model with expert prior knowledge for intelligent penetration testing. *Computers & Security*, 132:103358, 2023. URL https://www.sciencedirect.com/science/article/pii/S0167404823002687.

10 Hoang Viet Nguyen, Songpon Teerakanok, Atsuo Inomata, and Tetsutaro Uehara. The proposal of double agent architecture using actor-critic algorithm for penetration testing. In *ICISSP*, pages 440–449, 2021.

11 Carlos Sarraute, Olivier Buffet, and Jörg Hoffmann. Penetration testing == POMDP solving? *CoRR*, abs/1306.4714, 2013. URL http://arxiv.org/abs/1306.4714.

12 John Schulman, Filip Wolski, Prafulla Dhariwal, Alec Radford, and Oleg Klimov. Proximal policy optimization algorithms. *arXiv preprint arXiv:1707.06347*, 2017.

13 Cheng Wang, Akshay Kakkar, Chris Redino, Abdul Rahman, S Ajinsyam, Ryan Clark, Daniel Radke, Tyler Cody, Lanxiao Huang, and Edward Bowen. Discovering command and control (C2) channels using reinforcement learning (RL). *Submitted IEEE SouthEastCon 2023*, 2023.

14 Junkai Yi and Xiaoyan Liu. Deep reinforcement learning for intelligent penetration testing path design. *Applied Sciences*, 13(16):9467, 2023. URL https://www.mdpi.com/2076-3417/13/16/9467.

15 Mehdi Yousefi, Nhamo Mtetwa, Yan Zhang, and Huaglory Tianfield. A reinforcement learning approach for attack graph analysis. In *2018 17th IEEE International Conference On Trust, Security And Privacy In Computing And Communications/12th IEEE International Conference On Big Data Science And Engineering (TrustCom/BigDataSE)*, pages 212–217. IEEE, 2018.

16 Fabio Massimo Zennaro and Laszlo Erdodi. Modeling penetration testing with reinforcement learning using capture-the-flag challenges: trade-offs between model-free learning and a priori knowledge. *arXiv:2005.12632*, 2021.

17 Greg Conti and David Raymond. On cyber: towards an operational art for cyber conflict. *Kopidion Press*, 2018.

18 Shodan. (2024). The Shodan Search Engine. [Online]. Available: https://www.shodan.io/.

7

Putting it Into Practice: RL for Scalable Penetration Testing

7.1 Crown Jewels Analysis

7.1.1 Overview and Motivation

In the evolving landscape of cybersecurity, the protection of "crown jewels" (CJs) – the most critical assets within an organization's IT infrastructure – is paramount. This implementation delves into a specialized approach for crown jewel analysis (CJA) using reinforcement learning (RL), aptly named CJA-RL. Unlike broader cybersecurity strategies, this method zeros in on the unique characteristics and vulnerabilities of CJs. CJA-RL endeavors to automate the identification of crucial network components that are instrumental in safeguarding CJs. The objective is not just to locate potential weaknesses but also to understand the strategic importance of different network nodes in relation to the protection of CJs. This RL-based approach offers a nuanced perspective, focusing specifically on the dynamics of how CJs can be exposed to threats within the complex web of an organization's network.

The central motivation for this application is the disproportionate impact that security breaches can have on CJs compared to other parts of an organization's network. CJs, central to an organization's mission and operations, demand a targeted and sophisticated approach to security. Traditional cybersecurity methods, while comprehensive, often do not provide the specificity required to protect these vital assets effectively. This gap in cybersecurity methodology becomes more pronounced, considering that adversaries often target CJs not through advanced technical means but by strategically navigating and exploiting the network terrain. Thus, the need for a specialized approach like CJA-RL becomes critical. This method enhances our understanding of the network from the perspective of protecting CJs, offering insights into the most likely routes an

Reinforcement Learning for Cyber Operations: Applications of Artificial Intelligence for Penetration Testing, First Edition. Abdul Rahman, Christopher Redino, Dhruv Nandakumar, Tyler Cody, Sachin Shetty, and Dan Radke.

attacker might take and the key network nodes they might target to compromise these assets. The methodology's significance lies in its ability to pinpoint potential vulnerabilities and provide strategic insights into how these vulnerabilities, if exploited, could compromise an organization's most critical assets.

7.1.2 Network Setup for Evaluation

The evaluation of the CJA-RL method utilizes a network model meticulously designed to mirror the intricacies of a large-scale enterprise IT environment. The network's complexity is captured in an attack graph, comprising 1617 vertices and 4331 edges, making it one of the most extensive models used for RL-based cybersecurity testing. Constructed using MulVAL [4], the attack graph is not a simplistic representation of the network but an elaborate model that includes a variety of network components, such as servers, routers, and firewalls, along with different types of vulnerabilities and potential exploit paths.

Each vertex in the graph represents a distinct state in the network, covering a broad spectrum of network entities from individual nodes to complex configurations. The edges delineate potential actions an attacker might take, such as exploiting a vulnerability or moving laterally between network segments. This detailed graph is the foundation for the RL agent's decision-making, simulating a real-world network's complexity, and the myriad paths an attacker might navigate to compromise the CJs (Figure 7.1).

7.1.3 Reward Calculation

The reward system in CJA-RL is a pivotal aspect that directs the agent's learning process toward effective strategies. It incorporates several key elements:

- **Use of CVSS:** The common vulnerability scoring system (CVSS) plays a pivotal role in reward calculation. The transition probabilities and rewards are modeled based on the CVSS scores of exploits. For instance, transition probabilities are set at 0.9, 0.6, and 0.3 when moving out of a state with low, medium, and high

Algorithm 1 Crown Jewel Analysis via RL (CJA-RL)

Input: MDP, CJ node, reachable initial nodes
Output: Paths to CJ's 2-hop network and analysis

```
2-hop ← f_2-hop(CJ)                ▷ Get nodes in 2-hop network
for i in reachable initial nodes do
    for j in 2-hop do
        path ← f_RL(MDP, i, j)     ▷ Optimal path i → j
        paths ← store(path)
    end for
end for
return paths, analysis(paths)
```

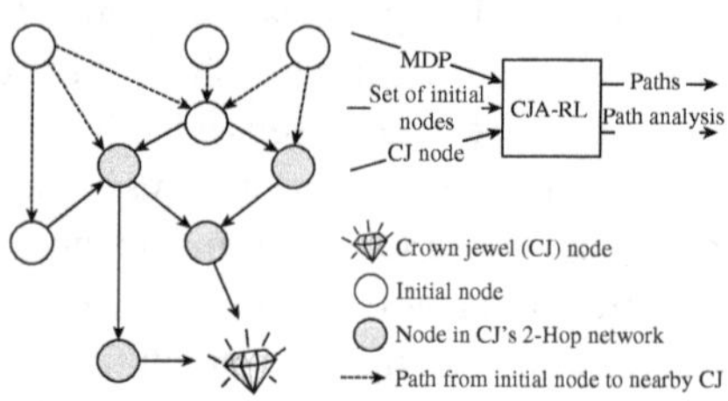

Figure 7.1 The CJA algorithm and network setup. Source: Adapted from Gangupantulu et al. [2].

attack complexity, respectively. The exploitability score of a state significantly influences the reward an agent receives for successful transitions.

- **Terminal Node Rewards:** Upon successfully arriving at a terminal node, which represents successful access to a CJ, the agent is awarded a high reward, typically set at 100. This is to emphasize the importance of reaching and compromising the CJs in the network.
- **Depth-First Search Path Rewards:** The agent is also rewarded for traversing nodes along a depth-first search path from the initial node to the terminal node. This reward is linearly scaled from 0 to 100, depending on the node's position in the path. This aspect of the reward system encourages the agent to follow efficient paths that minimize the number of hops while maximizing exploitability.
- **Incorporating Cyber Terrain:** Cyber terrain, especially the presence of firewalls as cyber obstacles, is factored into the reward mechanism. Transition probabilities are adjusted based on the type of protocols used (File Transfer Protocol, Simple Mail Transfer Protocol, Hypertext Transfer Protocol, Secure Shell), reflecting real-world network defense mechanisms. This inclusion enhances the realism of the simulation, preparing the RL model for practical application in actual network environments.

This sophisticated reward structure, combined with the detailed network model, positions CJA-RL as a powerful tool for simulating and understanding attack strategies in complex network environments, particularly in the pursuit of protecting critical assets like CJs.

7.1.4 Model Architecture

The CJA-RL model employs a deep q-network (DQN) approach, an advanced form of Q-learning utilizing deep neural networks (DNNs). The neural network, denoted as $Q(s, a; \theta)$, approximates the optimal action-value function, where θ represents the neural network parameters.

7.1.5 Training Process

The training of the CJA-RL model involves the following key steps:

1. **Construction of the MDP**: The Markov decision process (MDP) for the network is constructed using the attack graph generated by MulVAl. This graph's vertices represent the state space S, and its edges represent the actions A. Each vertex can be a network component or a traversal mechanism, while the edges signify the potential actions an attacker can take within the network.
2. **Defining Transition Probabilities and Rewards**: Transition probabilities $P(s, a, s')$ and rewards $R(s, a)$ are modeled based on the CVSS. Probabilities

are adjusted based on the attack complexity of the next state, and rewards are assigned based on exploitability scores and additional factors like reaching the terminal node or following a depth-first search path.

3. **Deep Q-Network Training**: The DQN is trained over multiple episodes, each representing a penetration testing scenario. The agent interacts with the MDP, taking actions based on its current policy and observing rewards and new states. The objective is to maximize the sum of discounted future rewards.
4. **Hyperparameters**: Key hyperparameters for the DQN include a learning rate for the optimizer, discount factor γ for future rewards, and parameters for the neural network architecture like the number of layers and neurons per layer. The exploration-exploitation trade-off is managed using an epsilon-greedy strategy.
5. **Policy Optimization**: The policy is iteratively updated based on the observed rewards and the predictions from the neural network. The goal is to converge to a policy that maximizes the cumulative reward, indicating effective paths and strategies for penetrating the network and reaching the CJs.

7.1.6 Experimental Results

In our work on the crown jewel analysis using reinforcement learning (CJA-RL), we identified the strategic importance of certain network hosts in proximity to CJs. Our approach highlighted that hosts within a 2-hop distance from a CJ, possessing vulnerabilities or exploitable elements, are crucial in sophisticated attack strategies. These hosts are essential stepping stones for launching advanced attacks aimed at compromising or extracting data from CJs, making them key focal points in offensive and defensive cybersecurity strategies.

In evaluating the effectiveness of our CJA-RL approach, we adopted a multi-dimensional method. Rather than solely concentrating on the rewards accrued by the agent, our evaluation provided an in-depth analysis of different CJs within a network. This comprehensive assessment approach offered a more practical understanding of CJA-RL's performance in real-world penetration testing scenarios, moving beyond theoretical models to actual network applicability.

A significant aspect of our work involved analyzing the network dynamics through the lens of key nodes in relation to various CJs. We identified the "best terminal nodes" as those requiring minimal hops to reach while maximizing rewards, positioning them as ideal for staging subsequent attacks on CJs. Furthermore, we pinpointed "most visited nodes" to highlight common pathways or critical junctures in the network, making them ideal targets for sensors or trackers. The analysis of "best initial nodes" also provided strategic insights for attack planning, revealing the most susceptible points for network entry.

Table 7.1 Experimental results from CJA RL.

Crown jewel	Best initial node	Best terminal node	Most visited node
AD computer – comp00384	Attack on abraniff00503 in comp00384	RuleHasActiveSession (cclinker00753)	RULE 1 (Exploit active session)
AD computer – fllabdc	Attack on lcopa00557 in fllabdc	Attack on misby0048 in fllabdc	RULE 1 (Exploit active session)
AD Group – administrators	Active Session exploited for Irainer00755	Attack on sfordham00591 in comp00001	RULE 28 (Users with active login sessions)
Object in subnet network_core	Attack on machine dbranstad00508	Root access to cloud_conn_mgmt	RULE 1 (Exploit active session)
Mail_server object in any subnet	Attack on efeltus00249 in comp00955	Attack on wfeagins00889	RULE 28 (Users with active login sessions)

Source: Adapted from Gangupantulu et al. [2].

In the following table, our findings reveal a notable tendency for the most visited nodes to be closely associated with active session interaction rules. This highlights their pivotal role in network traversal, especially for lateral movement within the network. As focal points for network traffic, these nodes provide strategic insights into potential attack paths. Additionally, terminal nodes often involve attacks on host computers within subnets. This pattern suggests a preference for engagement through interaction rules when attempting to access specific CJs. Such insights are vital for understanding the network dynamics and formulating attack and defense strategies (Table 7.1).

The insights gained from this application of CJA-RL have significant implications for cybersecurity. They provide a nuanced understanding of attackers' methods of discreetly navigating through networks, an essential knowledge for formulating effective defensive strategies. Furthermore, understanding these pivotal hosts' location and role is crucial for compromising a CJ and deploying effective countermeasures.

7.2 Discovering Exfiltration Paths

7.2.1 Overview and Motivation

The intricate landscape of cybersecurity necessitates a continuous evolution of defensive strategies. In this context, the unauthorized movement of data within a network, known as data exfiltration, remains a critical challenge. Adversaries often employ sophisticated techniques to move stolen data stealthily, circumventing conventional detection mechanisms. This application introduces a novel approach utilizing RL in conjunction with attack graphs and cyber terrain to identify optimal exfiltration paths in enterprise networks.

This research builds upon the previous section focused on identifying CJs within a network, reversing the scenario to assume that data has already been compromised and needs to be quietly exfiltrated. This work's unique contribution lies in the development of an RL-based reward function tailored to find paths where adversaries can minimize detection, a critical aspect of successful data exfiltration.

Data exfiltration is a two-pronged process involving both accessing the data and then transferring it undetected. Traditionally, the focus has been on preventing unauthorized entry into the network. However, moving data from within the network to an external location poses a significant challenge, often overlooked in cybersecurity strategies. This application addresses this gap by focusing on the latter part of the exfiltration process, leveraging RL to simulate and analyze potential exfiltration pathways. It is distinguished by its focus on modeling service-based defensive cyber terrain within dynamic attack graphs. We propose an RL-based algorithm to discover the top-N exfiltration paths in a network. This methodology aligns with the necessity of understanding network structure, configuration, and cyber terrain. It leverages scalable attack graph construction frameworks and integrates with existing vulnerability reporting systems like the CVSS.

7.2.2 Network Setup for Evaluation

The elements to consider for evaluating network setup are composed of various elements such as subnets, hosts, operating systems, privilege access levels, network services, and firewall rulesets. This setup was derived from architectural leading practices to accurately represent enterprise network configurations and deployments. The network is defined with the following properties.

- **Subnets:** Nine distinct subnets are defined, representing different segments of the network, such as server services, client workstations, demilitarized zones (DMZs), and core security services.
- **Hosts:** 26 hosts with varying configurations, including different operating systems and privilege levels.
- **Operating Systems:** Two types of operating systems, reflecting common diversity in enterprise environments.
- **Privilege Access Levels:** Three levels, mirroring real-world access control systems.
- **Network Services:** Nine different services, including database and server services.

- **Host Processes:** Six types, representing various operations running on the hosts.
- **Firewall Rulesets:** 39 rulesets, configuring the flow of data and access permissions within the network.

This environment was constructed to mimic common IT practices, focusing on realistic server, database, and client service configurations, alongside network security strategies like least-privilege and zero-trust models.

The environment is structured as follows.

- **State Representation:** Each host in the network is represented by a 1D vector detailing its status (compromised, reachable, discovered, or not) and configurations (address, services, operating system, and processes). The combined vectors of all hosts form a state tensor, describing the entire network.
- **Actions:** Defined as operations performed on a target host, including scanning, exploiting, or privilege escalation. The action must align with the target host's configuration for success.

7.2.3 Reward Calculation

The reward function is designed to simulate the risk and cost associated with different attacker actions within the network. It focuses on service-based risk penalties, enhancing the core CVSS-MDP model used in the CJA-RL approach as follows.

- **Service-Based Defensive Terrain:** Rewards are based on the assumption of defenses being present in correlation with the services available on a host. This includes antivirus software, firewalls, and authentication tracking.
- **Service Categories and Action Types:** The model categorizes services into authentication, data, security, and common services, with different penalties for exploiting these services.
- **Negative Reward Structure:** A quantified negative reward system is used, with penalties for actions based on service categories and action types. When performing an exploiting action, this hierarchy applies authentication as a reward of −6, data as a reward of −4, while security and common have a reward of −2. When performing a scanning action, the reward is increased by 1 (i.e., −5, −3, −1).
- **Risk Hierarchy and Penalties:** Penalties are assigned based on the perceived risk associated with each service category, aligning with real-world prioritization of security controls.
- **Realistic Attack Simulation:** The approach models an attacker's decision-making process, considering the risk of detection and the cost of actions on different services.

When an action is taken on a host with multiple services, the highest cost (most negative reward) associated with any of the services present on the host is applied. This decision is based on a "leading practice" approach in cybersecurity, where security controls are applied based on the riskiest service on a host.

7.2.4 Model Architecture

The architecture leverages a sophisticated implementation of the advantage actor-critic (A2C) algorithm, chosen for its balance between exploration and exploitation, which is crucial in network security's dynamic and uncertain landscape. The A2C model's efficiency in policy learning is augmented by its DNNs, featuring layers with specific sizes and activation functions. These neural networks use tanh activation functions for non-output layers and a softmax function for the output layer, optimizing the model's ability to process and learn from complex network environments.

Complementing the A2C model is the double agent (DA) architecture, inspired by Nguyen et al., which employs two A2C agents working in tandem. This dual-agent approach is designed to divide the task of network penetration into two distinct phases, thereby providing a more holistic and comprehensive analysis. The first agent is responsible for iteratively scanning the network and building a structural model, while the second agent exploits this model to discover optimal exfiltration paths. This division of labor allows for a more detailed and nuanced approach to network penetration, mirroring the complexities encountered in real-world scenarios.

Hyperparameters play a pivotal role in tuning the models for optimal performance. Key hyperparameters include the learning rate, set at 0.001, and the discount factor, set at 0.99, which govern the model's learning speed and the valuation of future rewards. The models are trained for 4000 episodes, each with a maximum of 3000 steps, ensuring thorough exploration and learning within the simulated network environment. The A2C model and the structuring agent of the DA use DNNs with layers sized 64, 32, and 1, whereas the exploiting agent of the DA uses layers sized 10 and 1, tailored to their specific roles within the architecture.

Risk penalties are an integral component of the model, introducing a layer of realism by simulating the varying degrees of risk associated with different network services and actions. These penalties are scaled to represent different risk preferences: risk-averse, risk-neutral, and risk-accepting, with values of 1.3, 1.0, and 0.7, respectively. This scaling affects both the convergence of the agents and the paths they discover, reflecting the strategic considerations an attacker would weigh in a real-world scenario.

The combination of the A2C model, the DA architecture, the meticulously chosen hyperparameters, and the nuanced risk penalties culminate in a robust

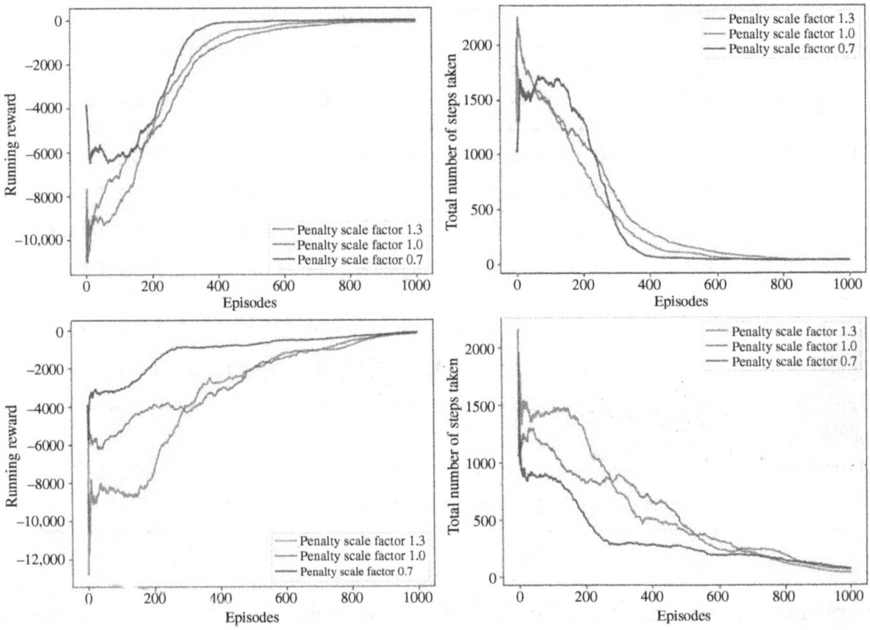

Figure 7.2 Training metrics and plots for exfiltration path discovery. Source: From Cody et al. [1], 2022, IEEE.

and realistic simulation of network exfiltration scenarios. The architecture and algorithms are tailored to adapt and learn in an environment characterized by uncertainty and variability, mirroring the challenges faced in real-world network security. This comprehensive approach enhances the model's effectiveness in identifying exfiltration paths and provides valuable insights and tools for cybersecurity professionals engaged in network defense and penetration testing (Figure 7.2).

7.2.5 Experimental Results

The experimental results provide a detailed and insightful examination of the RL models' capabilities in simulating network exfiltration paths. These results are not only crucial in validating the effectiveness of the models but also in offering practical insights for cybersecurity applications (Table 7.2).

The convergence of the models within a thousand episodes is a significant finding, demonstrating their ability to learn and adapt to a simulated network environment efficiently. The DA architecture, with its inherent complexity due to the interaction between two A2C models, showed a slower convergence rate compared to the standalone A2C model. This difference in convergence rates is not a mere

Table 7.2 Experimental results for exfiltration path discovers.

Path rank	Scale factor	Path	Steps	Reward	Cumulative probability score
Best path	0.7	(6, 0) → (3, 0) → (2, 0)	12	57.8	2.9 + 2.9 = 5.8
	1.0	(6, 0) → (3, 2) → (1, 0)	11	62	2.9 + 2.9 = 5.8
	1.3	(6, 0) → (3, 0) → (1, 0)	5	68.3	1.9 + 2.9 = 4.8
Second best path	0.7	(6, 0) → (3, 2) → (1, 0)	19	46.9	2.9 + 4.9 = 7.8
	1.0	(6, 0) → (3, 0) → (2, 0)	16	24	2.9 + 4.8 = 7.7
	1.3	(6, 0) → (3, 2) → (1, 0)	19	33.1	1.9 + 2.9 = 4.8
Third best path	0.7	(6, 0) → (3, 0) → (1, 1) → (4, 0)	15	41.3	1.9 + 1.9 + 2.4 = 6.2
	1.0	(6, 0) → (3, 2) → (1, 0) → (4, 0)	24	17	3.9 + 1.9 + 7.5 = 13.3
	1.3	(6, 0) → (3, 2) → (1, 0) → (4, 0)	22	−6.1	1.9 + 2.9 + 6.3 = 11.1

Source: From Cody et al. [1], 2022, IEEE.

performance metric but reflects the depth of analysis conducted by the DA model, making it particularly suited for more complex network environments where a layered approach to penetration testing is necessary.

The agent behavior under various risk preferences – risk-averse, risk-neutral, and risk-accepting – revealed the adaptability of the models to different strategic considerations. This adaptability is crucial in real-world applications where attackers may have different risk tolerances. The paths chosen by the agents varied with the risk settings, indicating how these preferences could influence an attacker's strategy in navigating a network for data exfiltration. This has significant implications for cybersecurity defenses, as it assists in anticipating and preparing for various attack strategies.

Furthermore, the agents' ability to identify and exploit network misconfigurations in the simulation indicates their potential in discovering real-world vulnerabilities. This is particularly important for enhancing network security, as it shows the model's effectiveness in identifying potential exfiltration paths and vulnerabilities within a network.

The results are invaluable for cybersecurity professionals, providing insights into potential exfiltration paths, and strategies an attacker might employ. This knowledge is crucial in strengthening network defenses and developing more effective mitigation strategies. The variations in path selection under different risk profiles assist in strategic defense planning, enabling cybersecurity teams to prioritize defenses based on the most likely paths an attacker would take, given their risk profile.

As for the payload, we acknowledge that the current study does not account for the size or rate of the data payload being exfiltrated. While incorporating these variables would increase the complexity of the model, we recognize their importance in real-world data exfiltration scenarios. Upcoming sections of this chapter will explore these considerations.

7.3 Discovering Command and Control Channels

7.3.1 Overview and Motivation

The primary objective of this application is to advance the discovery of C2 channels using an RL-based approach. This method aims to automate the process of carrying out C2 attack campaigns in large-scale network environments that incorporate multiple layers of defense. By modeling C2 traffic flow as a three-stage process and formulating it as a MDP, the study seeks to maximize the number of valuable hosts from which data is exfiltrated. A novel aspect of this approach is its specific modeling of payload and defense mechanisms, such as firewalls.

The motivation behind this study is twofold. First, it addresses the challenge of detecting sophisticated cyber actors, particularly advanced persistent threats (APTs), which architect C2 channels within networks to maintain long-term footholds and avoid detection. Second, the study proposes an efficient and automated way to identify potential attack pathways, which traditionally is a time-consuming and complex task when done manually.

The approach develops a detailed RL model that incorporates cyber defense terrain for staging multi-phase C2 attacks, a notable advancement in the field. Furthermore, it demonstrates the model's effectiveness in identifying attack paths within a large-scale network, constructed using real-world data, encompassing over 100 subnets and 1400 hosts. This expansive network serves as a testing ground to validate the RL model's ability to navigate complex network environments and identify potential C2 channels effectively.

7.3.2 Network Setup for Evaluation

The environment for the C2 channel discovery study using RL is meticulously constructed to simulate real-world network conditions, incorporating a complex structure of communication paths, routers with varying security levels, and endpoints with potential vulnerabilities. This environment is programmatically created, drawing upon real-world experiences from penetration testers and security analysts, and is designed to accommodate the implementation of a three-stage campaign model.

7.3.2.1 Three-Stage C2 Attack Model

The C2 attack campaign in this study is modeled as a three-stage process: infection, connection, and exfiltration. Initially, the attacker (or the RL agent in this context) aims to gain a foothold in the network by exploiting known vulnerabilities on target hosts. After successful infiltration, the agent seeks to establish communication with the C2 server for further instructions. This could involve locking or sending out specific files from the compromised system. The final stage involves exfiltrating valuable information from the target system to the remote server, marking the completion of the attack.

7.3.2.2 Network Exploration and Exploitation

During the infection phase, the RL agent can perform subnet scans or exploit actions on targeted hosts. Subnet scans reveal all hosts on the same subnet and those with certain services on adjacent subnets. Each exploit action is tied to a specific common vulnerabilities and exposures (CVE) vulnerability, and its success depends on the presence of a particular service, process, and the operating system of the target host.

7.3.2.3 Connection and Exfiltration Phases

Once a sensitive host is compromised, the agent can initiate connection attempts to the C2 server. To establish a connection, the traffic must navigate through all firewalls between the host's subnet and the Internet. Connection attempts might be blocked if any firewall has undergone an update since the host's infection, and too frequent connection attempts can raise alerts, triggering an emergency firewall update. The exfiltration phase involves identifying a target payload size and uploading portions of it over time, requiring strategic pauses to avoid detection by firewalls.

7.3.2.4 Firewall Dynamics

Firewalls play a critical role in the network environment, located between all subnets and the Internet. They are updated periodically or in response to unusual traffic patterns, with a wall-clock time mechanism determining the schedule for regular updates. Emergency updates are triggered by specific conditions like excessive connection attempts or abnormal upload volumes.

7.3.2.5 RL Formulation - State Space, Action Space, and Reward Function

The state space in the RL formulation includes features for each host, such as subnet and local ID, operating system, services and processes, and discovery and infection values and statuses. The action space comprises subnet scans, exploits, connection attempts, uploads, and sleep actions across the three attack stages. The reward function is dual-part: it rewards progress toward the goal and imposes a cost for specific actions, which may trigger the network's defense systems. Positive rewards are granted for successfully discovering, exploiting, and connecting a host, and for uploading payloads. A significant bonus is awarded upon completing the entire payload upload. However, detection by firewalls results in penalties, negating all accumulated rewards and isolating the host.

7.3.3 Model Architecture and Training

The training of the RL agent is a pivotal component, executed with precise methodologies and hyperparameters to ensure effective learning in a complex network environment.

7.3.3.1 Training Methodology and Network Architecture

The RL agent is trained episodically using the proximal policy optimization (PPO) algorithm, a choice driven by its balance between exploration and exploitation and its suitability for complex tasks like C2 channel discovery. Each training episode concludes when all sensitive hosts either successfully send their payload to the C2 server or become isolated by firewalls. This approach simulates a realistic attack scenario where the agent must navigate through network defenses to achieve its objective. To prevent excessively long episodes, the maximum number of time steps per episode is capped at 10,000.

The neural network architecture for both the actor and the critic involves a two-layer feed-forward design. The first layer consists of 128 neurons, and the second layer contains 64 neurons. This configuration allows the model to effectively process the intricate patterns and strategies involved in identifying and utilizing C2 channels within the network.

7.3.3.2 Hyperparameters

The selection of hyperparameters is critical to the training process. The critic learning rate and actor learning rate are both set at 0.0003 and 0.00003, respectively, ensuring a stable and gradual learning curve. A high discount factor of 0.99 indicates the agent's strong consideration for future rewards, encouraging it to plan for long-term outcomes rather than focusing solely on immediate gains.

The horizon, or the number of steps in each episode before the gradients are computed, is set at 4096. This provides the agent with a substantial experience window within each episode to learn from. The minibatch size is chosen as 64, and the model is trained over 5 epochs, allowing for multiple passes over the training data for better generalization. The generalized advantage estimation (GAE) parameter is set at 0.95, aiding in reducing the variance of the policy gradient estimates. The clipping parameter at 0.2 is utilized to prevent overly large updates to the policy, thereby maintaining training stability. Finally, an entropy coefficient of 0.001 is included to encourage exploration by the agent.

7.3.3.3 Training Execution and Computational Resources

The RL agent undergoes extensive training, with the process spanning five million iterations. This extensive training is essential to ensure the agent comprehensively learns and adapts to the various strategies required for successful C2 channel identification. To validate the model's robustness and consistency, the training process is repeated five times, each with different random seeds. This repetition ensures the reliability and generalizability of the learned policies.

All experiments are conducted on high-end computational resources, specifically two Intel Xeon Platinum 8124M processors, each with 18 cores. The use of such powerful computing resources underscores the computational intensity and complexity of the training process, a necessity given the intricate nature of the network environment and the task at hand.

7.3.4 Experimental Results

The experimental results from the study on discovering C2 channels using RL demonstrate the RL agent's ability to efficiently navigate and complete tasks within a large network environment.

7.3.4.1 Training and Convergence

The training process proved stable, with the RL policy converging over 10,000 episodes. A significant outcome was the steady increase in the sum of rewards per episode, reaching close to 26,000, which is near the theoretical maximum under the established reward structure. Concurrently, the average episode length gradually decreased, eventually averaging just over 100 steps. This reduction in episode length indicates that as training progressed, the RL agent became more efficient in completing the attack task, minimizing random or unnecessary actions (Figure 7.3).

7.3.4.2 Evaluation of Learned Policy

To evaluate the final learned policy, 100 attack paths were sampled using the trained actor-network. In 78 of these trajectories, both target hosts completed sending payloads to the C2 server. In the remaining 22 scenarios, one of the target hosts was blocked by firewalls. However, in all cases, at least one target host was successfully attacked by the RL agent. The agent typically completed the task in 107 steps or 47 minutes, with an average total reward of 25,824.

7.3.4.3 Behavioral Analysis of RL Agent

The RL agent, due to the stochastic nature of the learned policy, occasionally took unnecessary actions, such as exploiting unimportant hosts or redundantly connecting a host to the C2 server. After analyzing the best-performing trajectory, key

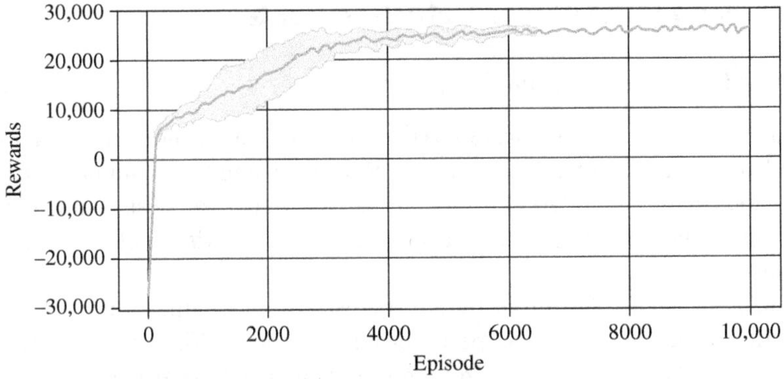

Figure 7.3 Reward accrual across episodes for command and control channel discovers. Source: From Wang et al. [5], 2023 IEEE.

steps in the C2 attack process were identified. For example, the agent consistently chose the fast upload option over the slow one, indicating a strategic understanding of when to pause or sleep to evade detection. On average, it took the agent 4.2 attempts to establish connections to the C2 server, with at least 20 upload actions needed for a successful exfiltration of a 20,000 MB payload across two targets. The agent also averaged over 32 deliberate sleep actions in each episode to mask its activities.

7.3.4.4 Avoidance of Firewall Detection

A critical aspect of the RL agent's strategy was its ability to circumvent firewall detection during uploads. The agent uploaded data at a regular cadence, a tactic that can be analyzed by security experts to develop more sophisticated defense strategies. The pattern of upload actions revealed that the agent typically waited for about two minutes, equivalent to taking two sleep actions, before resuming the upload process. This strategy effectively utilized available bandwidth for data exfiltration while keeping the total upload volume within monitored windows well below alert thresholds (Figure 7.4).

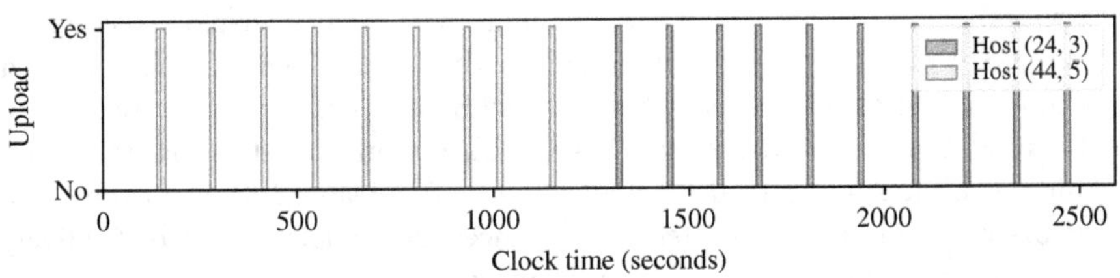

Figure 7.4 Times of upload actions taken by agent. Source: From Wang et al. [5], 2023 IEEE.

7.4 Exposing Surveillance Detection Routes

7.4.1 Overview and Motivation

The primary objective of this application is to develop and refine methods for identifying and analyzing SDRs within enterprise networks using RL techniques. This research stands on the shoulders of prior efforts in RL for path analysis, but with a distinct focus on the creation of SDRs that emphasize evading risk while exploring network services.

SDRs are critical pathways used by adversaries to gather intelligence about targets within a network, such as ports, protocols, applications, and services. Understanding and exposing these routes is essential for cybersecurity, as it enables defenders to anticipate and mitigate potential threats. This work aims to harness the power of RL in combination with attack graphs and cyber terrain principles to establish a robust framework for this purpose.

The motivation for this research stems from the ever-evolving landscape of cyber threats and the need for advanced methods to detect and counteract adversaries' sophisticated reconnaissance activities. Traditional cybersecurity measures often rely heavily on human analysis and standard automated systems, both of which have limitations in terms of scalability, speed, and depth of analysis. By applying RL to this domain, the researchers seek to transcend these limitations, offering a more dynamic and effective approach to identifying potential surveillance routes used by attackers.

The work is also driven by a desire to enhance the automated detection capabilities within cybersecurity infrastructures. The integration of RL with current security information (network topology and configuration) promises to highlight vulnerabilities and blind spots in existing security systems, paving the way for more proactive and preemptive defense strategies.

Notably, there are two distinct additions to this application that were not used in prior sections.

- **Warm-Up Phase in RL Algorithm:** A unique aspect of the RL algorithm used in this study is the incorporation of a 'warm-up' phase. This phase enables initial exploration and assessment of network areas that are safe to explore, based on a tailored reward and penalty system. This method mimics the cautious approach a human operator might take when conducting network reconnaissance.
- **Double Agent Architecture (DAA) Extension:** Building on the DAA approach, the study extends its application with advanced RL algorithms like actor-critic and PPO. This extension enhances the model's ability to effectively navigate and analyze complex network environments.

7.4.2 Network Setup for Evaluation

The environment for the SDRs is modeled as a MDP, a standard approach in RL. This model incorporates the complexities of a network environment, including hosts, services, and defensive mechanisms. The description for the network, hosts, and action spaces are detailed as follows.

- **Network Description:** The network is structured with multiple layers of defenses, such as firewalls and VPN management firewalls, to simulate real-world network conditions. This includes systems like HTTP and email services in the DMZ, private internal networks with controlled access, and various subnets separated based on access rules.
- **Host Representation:** Each host in the network is represented as a 1D vector encoding its status (compromised, reachable, discovered, or not) and configurations (address, services, OS, and processes). The collective vectors of all hosts form the state tensor of the environment.
- **Action Space:** The action space for the RL agent includes six primitive actions – scanning, exploiting, and privilege escalation.

All experiments start with one initial host, from where the agent must surveil the network. Details of the initial and target hosts, as well as the reward for terminal states were set as:

- **Initial Host:** The initial state of the environment is set with a compromised, reachable, and discovered host, acting as the starting point for the RL agent. This setup assumes that the attacker (the agent) already has a foothold in the network.
- **Terminal Hosts:** The terminal states are predefined as specific hosts within the network, with coordinates like (3, 1), (8, 2), or (9, 2). The goal is for the RL agent to explore all the services of these target hosts. A high positive reward (100) is assigned if the agent successfully reaches the terminal state and explores all the services of the target host. This incentivizes the agent to achieve the specified goal.

7.4.3 The Warm-Up Phase

The warm-up phase in the RL framework for SDRs serves as a critical prelude to the primary training phase. It is designed to preliminarily explore the network environment, identifying areas that are safe for further exploration. This phase is instrumental in setting the stage for more efficient and focused learning during the subsequent training phase.

Here's how the warm-up works:

- **Initial Exploration:** During the warm-up phase, the RL agent interacts with the network in a series of episodes. Unlike the training phase, the agent in the warm-up phase does not update its learning weights. Instead, it observes and records the positive rewards obtained from different nodes in the network.
- **Positive Reward Monitoring:** The agent monitors for positive rewards, particularly from scanning actions. If a positive reward is received from a node, it indicates that the defense terrain permits access to this area, signaling a lower risk for exploration.
- **Node Selection and Goal Updating:** A crucial part of the warm-up phase involves updating the agent's goal. The algorithm allows for the inclusion of only one node per subnet, based on the maximum positive reward, into the agent's objectives. This mirrors the real-world strategy where controlling one node in a subnet often provides sufficient access to the entire subnet.
- **Formation of Dynamic Nodes:** At the end of the warm-up phase, the algorithm identifies a set of "dynamic nodes," along with the initial target node, which the agent must explore to achieve its objectives. The inclusion of these nodes is dynamic and changes based on the rewards observed during the warm-up episodes.

The warm-up is important for several key reasons as it enhances the efficiency of the agent's exploration but also sets a strong foundation for more advanced and targeted learning in a complex environment. Some additional details of the value of the warm-up phase are listed below.

- **Risk-Aware Exploration:** The warm-up phase introduces an initial, cautious exploration step that mimics human reconnaissance behaviors. This step is critical in RL, particularly in adversarial settings like penetration testing, where blind exploration could lead to high-risk encounters or detection.
- **Efficient Learning Pathway:** By identifying areas of the network that are relatively safe and rewarding to explore, the warm-up phase guides the agent to focus its learning on these parts. This targeted approach improves the overall efficiency and effectiveness of the learning process.
- **Adaptation to Network Environment:** The dynamic nature of the warm-up phase allows the RL agent to adapt its strategy based on the specific configuration and security posture of the network. This makes the RL approach more robust and applicable to a variety of network environments.
- **Foundation for Advanced Learning Strategies:** The information gathered during the warm-up phase forms the foundation upon which more complex learning strategies, like actor-critic methods and PPO, can be effectively implemented in the subsequent training phase.

7.4.4 Model Architectures and Training

Here, the experiment is conducted using a single-agent A2C model along with two variants of the DA architecture. These models are intricately designed to cater to the specific needs of network penetration and reconnaissance. The A2C model, a well-regarded algorithm in RL, forms the basis of the single-agent approach. Its architecture comprises a DNN with three fully connected layers of sizes 64, 32, and 1, reflecting a standard approach for balance in exploration and exploitation.

The DA models, on the other hand, add complexity and depth to this exploration. The original DA-A2C model has both its structuring and exploiting agents trained using the A2C algorithm. This is further enhanced in the DA-PPO model, where PPO is combined with the DA architecture. The integration of PPO is particularly notable for its improved sample efficiency and training stability, a crucial factor in the dynamic environment of network security. In these DA models, while the structuring agent uses a DNN similar to the A2C model, the exploiting agent employs a slightly simpler DNN with two fully connected layers of sizes 10 and 1. The choice of tanh activation functions for non-output layers and softmax for the output layer in all these DNNs underscores the focus on optimizing the agent's decision-making capabilities.

The training regimen of these models is meticulously set up to ensure comprehensive learning. Adam optimizer, known for its effectiveness in handling sparse gradients and adaptive learning rates, is used across all models. The A2C model undergoes training for 4000 episodes, while the more complex DA models are trained for double the duration, 8000 episodes, reflecting the added intricacies these models handle. Each episode is capped at a maximum of 3000 steps, with termination conditions tied either to the achievement of the goal – exploring all services of the target host, or reaching the step limit. This ensures a focused and goal-oriented training process.

Furthermore, the sensitivity analysis conducted through the application of various scale values – 1, 3, 5, 11, and an additional 15 for certain hosts, provides insights into the algorithm's performance across different levels of network exploration. These scale values, experimental in nature, serve as a measure of the agent's exploration depth in the network, with higher values denoting a more cautious approach. The training of the models to convergence is critical in this context, as it guarantees that the agents thoroughly learn and adapt to the optimal strategies for surveillance detection within the complex simulated network environment.

Through this detailed and methodical setup of model architectures, hyperparameters, and training procedures, we ensure a robust and effective approach to RL application in SDRs, effectively training the models to navigate and analyze enterprise networks for security vulnerabilities.

7.4.5 Experimental Results

The DA-PPO model demonstrated a remarkable convergence speed compared to the DA-A2C model across all penalty scales. It was observed that DA-PPO converged in less than ten minutes, a significant improvement over the 30 minutes required by the DA-A2C model. Additionally, DA-PPO required fewer episodes to learn an effective policy. This rapid training and convergence highlight the efficiency of the DA-PPO model, making it a suitable choice for real-world applications where time and computational resources are critical factors (Figure 7.5).

Furthermore, the experiments revealed a distinct pattern in pathfinding across different penalty scales. The models were tested with penalty scales ranging from 1 to 15, where a higher scale indicated a more risk-averse behavior. The results showed that as the penalty scale increased, the RL agent reduced the amount of exploration, focusing only on areas deemed safe. This behavior was consistent across various target hosts, indicating the model's adaptability to different network environments and its capability to minimize crossing defense lines.

The A2C model, being simpler, performed well at penalty scales of 1, 11, and 15, but showed limitations at scales of 3 and 5 for certain target hosts, suggesting its inadequacy in more complex or sensitive network scenarios. This finding is particularly relevant for cybersecurity applications where the accuracy and reliability of pathfinding are crucial.

The experimental setup and results also provide insights into how RL agents can emulate human actors in cybersecurity scenarios. For example, at a penalty scale of 1, the agent's behavior is akin to actors in "smash and grab" operations or those with less experience, who do not prioritize stealth and are willing to accept higher risks. These agents were observed to perform more "noisy" scans, involving enumeration of multiple network services and traversing various network segments, potentially increasing their detectability.

In contrast, at a penalty scale of 11, representing highly-competent actors like nation-state actors or APTs, the agent displayed highly risk-averse behavior. It opted for the most direct paths and avoided traversing multiple protocols or services, thereby minimizing its footprint and chances of detection. This behavior mirrors the strategic approach of experienced cyber actors who prioritize stealth and minimize risk (Figure 7.6).

The experimental results underscore the capabilities of RL models, particularly DA-PPO, in efficiently navigating complex network environments while adapting to varying levels of risk aversion. These models' ability to mimic human-like decision-making processes in cybersecurity contexts, coupled with their efficiency in training and convergence, demonstrates their potential for practical applications in network security and penetration testing. These results not only validate

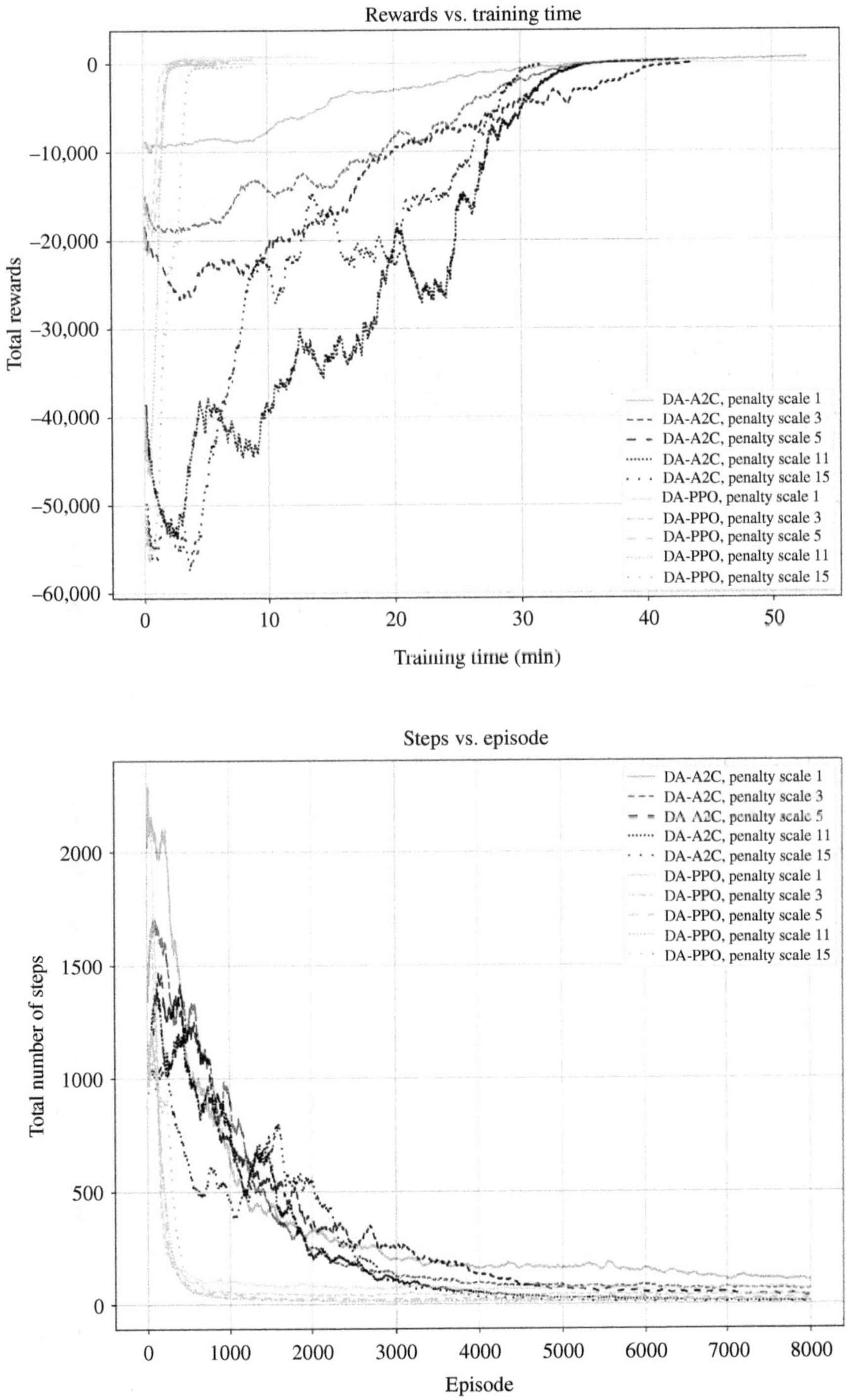

Figure 7.5 Training plots for SDR path analysis. Source: From Huang et al. [3], 2022, IEEE.

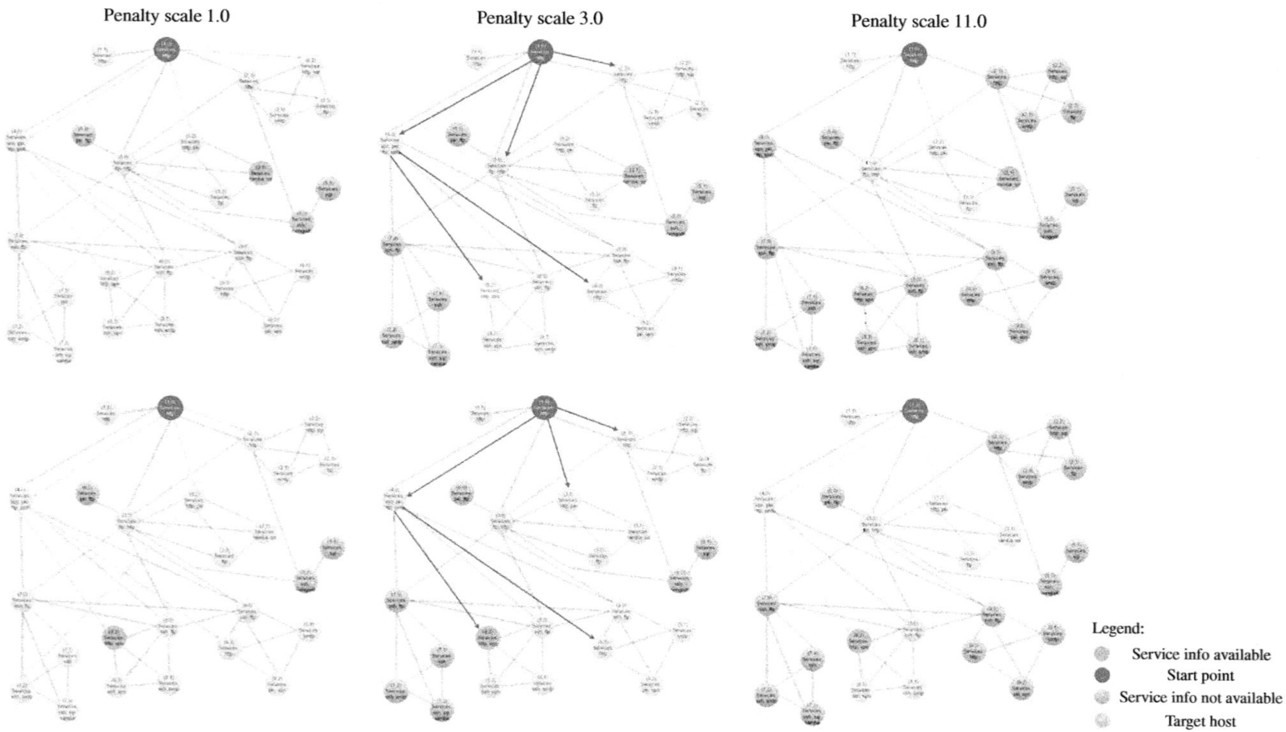

Figure 7.6 SDR paths at varying penalty scales between models. Source: From Huang et al. [3], 2022, IEEE.

the effectiveness of the proposed RL models in SDR but also offer valuable insights for their deployment in real-world cybersecurity scenarios.

7.5 Enhanced Exfiltration Path Analysis

7.5.1 Overview and Motivation

In this section, we delve deeper into the realm of data exfiltration, building upon the concepts introduced earlier in the chapter. While the prior section laid the groundwork by utilizing RL for identifying exfiltration paths, this segment significantly expands upon that methodology. Here, we introduce the critical elements of protocol and payload considerations, aspects that were not fully explored previously. This advancement is pivotal in more accurately simulating the nuances of network-based exfiltration events, particularly in the context of complex adversarial behaviors such as the exfiltration of large payloads over time or using specific protocols to evade detection.

The initial exploration into exfiltration path analysis, as discussed earlier, was primarily focused on optimizing paths within network topologies through a specialized reward system. However, it primarily catered to scenarios with nominal payload sizes and did not factor in the intricacies of payload size or protocol preferences during exfiltration operations. Recognizing these limitations is crucial, especially when considering large-volume exfiltration operations, which are more reflective of real-world cyber security paradigms. Adversarial activities in such contexts often involve choosing specific protocols, like tunneling exfil traffic through domain name systems (DNS), to effectively mask intentions and avoid detection.

In this section, we address these shortcomings by integrating communication payload and protocol preferences into the RL framework. This enhanced approach allows for a more nuanced emulation of adversarial considerations, such as managing the size and rate of data exfiltration, and choosing protocols that align with stealth and long-term data theft strategies. We now include a range of protocol choices and payload sizes, providing a more comprehensive model of how adversaries might behave in a network exfiltration context.

Aligned with the detailed examination of network structure, path analysis configurations, and cyber terrain, this methodology offers a clearer understanding of potential exfiltration routes. Emphasizing reproducibility and thoroughness, the RL solution methods, experimental design, and network model are articulated in great detail.

The introduction of this enhanced methodology is significant for its depth and realism. It builds upon the foundational concepts about exfiltration introduced earlier, providing a more accurate tool for practitioners in anticipating and understanding complex adversarial behavior in cyber environments. By focusing on payload and protocol preferences, this section marks a crucial advancement in the application of RL for cybersecurity, particularly concerning post-exploitation activities like data exfiltration.

7.5.2 Network Setup for Evaluation

The environment is structured with a focus on realism, particularly in terms of network firewalls and protocol-based path selection, to accurately simulate an attacker's campaign.

7.5.2.1 Exfiltration Campaign Model

The exfiltration campaign is modeled around three fundamental tasks: connection, path selection, and exfiltration. The process begins with the attacker gaining control over target hosts that are externally connected to the internet, serving as potential points of exfiltration. Once a host is compromised, the attacker, or the RL agent in this case, selects an exfiltration path based on protocol preferences. The agent then starts exfiltrating data packets from the compromised host. This dynamic model allows for the adjustment of exfiltration paths in real-time if the agent discovers a new, more suitable exfiltration host.

7.5.2.2 Network Firewalls and Monitoring

Network firewalls play a crucial role in the simulation. Positioned between each of the subnets and the public Internet, they are tasked with monitoring all exfiltration traffic. The system is designed to alert administrators upon detecting unusual traffic patterns that exceed predefined thresholds. For instance, if the total egress volume surpasses the maximum upload volume or if the active time exceeds the maximum upload duration, an alert is triggered, followed by an emergency firewall update. This update includes patching vulnerabilities and blocking outbound traffic from compromised hosts. To add to the realism, the simulation includes a wall-clock mechanism to emulate the real-time duration of an attack. This feature adjusts the clock time variably based on the complexity of different actions the RL agent takes, thereby simulating the actual passage of time in an attack scenario.

7.5.2.3 Protocol-Based Path Selection

The selection of exfiltration paths is heavily influenced by the use of common protocols, such as HTTPS. These protocols are chosen because they are generally deemed safer and less likely to be detected by security monitoring systems. By utilizing protocols commonly used in enterprise applications, the model emulates real-world attacker behaviors where they exploit these protocols to avoid raising alerts in monitoring systems. The path selection algorithm in the model is designed to maximize the utilization of the chosen protocol across as many hosts as possible in the exfiltration network path. The algorithm prioritizes finding a complete path using the designated protocol over simply choosing the shortest path available. In scenarios where a complete path using the preferred protocol is not possible, the algorithm seeks out the shortest path that maximizes the use of the protocol. The reward function plays a significant role here, as it calculates the most rewarded path based on the shortest number of hosts and maximum use of the exfiltration protocol.

7.5.2.4 Action Clock Time and Reward Function

The simulation also considers the time taken for each action, which varies based on the action's result and complexity. This detail adds another layer of authenticity to the simulation, as different actions in a real-world attack would naturally take varying amounts of time. The reward function is another critical component, offering positive values for achieving sub-goals like discovering or exploiting a host and negative values accounting for the action's cost. The cost of an action is likely to trigger the defense terrain, and it is calculated based on the services running on the target system. The services are categorized into high-risk, medium-risk, and low-risk groups, and the actual cost of an action depends on its type and the target's service profile.

7.5.2.5 Training and Evaluation Networks

Two distinct experimental networks are utilized to test and refine the RL agent's capabilities. Each network is intricately designed to simulate different complexities and scenarios that an attacker might encounter in real-world cyber environments.

The first network is relatively smaller, comprising 10 subnets with a total of 56 hosts. The distribution of hosts across these subnets varies, with each subnet containing between 3 and 12 hosts. This network is designed to simulate a more constrained and less complex environment, allowing for the initial testing and adaptation of the RL agent's strategies. The RL agent is assumed to have already compromised a host within this network, specifically host (8 2) in subnet 8, which is not directly connected to the Internet. This setup introduces a realistic challenge for the agent: to navigate through the network from an initial foothold that is not straightforwardly accessible from the external network.

A key element in this first network is the designation of a specific host, (2 0) in subnet 2, as the exfiltration host. This subnet has the unique feature of being directly accessible from the internet, unlike the other private subnets. The exfiltration host in this subnet runs the dynamic host configuration protocol server (DHCPS) service, chosen as the exfiltration protocol for this network. The selection of DHCPS for exfiltration aligns with realistic attacker behaviors, where commonly used protocols are exploited to blend in with normal network traffic and avoid detection.

The second network represents a significantly larger and more complex environment, consisting of 101 subnets with a total of 1444 hosts. This expansive network is designed to test the RL agent's capabilities in a more intricate and challenging scenario, closely mimicking large enterprise networks. The hosts in this network are more varied, with each subnet containing between 3 and 50 hosts.

In this network, the RL agent is also assumed to have gained an initial foothold, this time on host (44 5) in subnet 44, which, similar to the first network, is not directly connected to the Internet. The designated exfiltration host is located in subnet 5, specifically host (5 10), which is connected to the internet. The service running on this exfiltration host is HTTPS, chosen as the Exfiltration Protocol. The use of HTTPS, a standard and commonly used protocol, adds another layer of realism to the simulation. It reflects an attacker's preference for protocols that are less likely to raise alarms within network monitoring systems.

These two networks, with their differing scales and complexities, provide a comprehensive environment for evaluating the RL agent's ability to identify and utilize exfiltration paths effectively. The variation in network size, host distribution, and chosen exfiltration protocols between the two networks allows for a robust assessment of the RL model's scalability and adaptability in diverse cyber terrain.

7.5.2.6 State and Action Spaces

The state space in the model is detailed for each host, including address, operating system, services and processes, discovery status, infection status, and access level information. For target hosts, additional features like connection status, time since infection, and remaining payload size are included. The RL agent's action space comprises subnet scans, exploits, uploads, and a sleep action for periods of inactivity. Multiple exploit options targeting different vulnerabilities are available for each host, along with upload actions at varying speeds.

7.5.3 Model Architecture and Training

The RL agent's training is an integral part of the methodology, involving a sophisticated setup of model architectures, hyperparameters, and a structured training regimen.

7.5.3.1 Model Architecture and Training Approach

The study employs the PPO algorithm for training the RL agent, a well-known choice in RL for its balance between exploration and exploitation. This algorithm is particularly suited for tasks with significant complexities, like the one at hand, involving data exfiltration path analysis in network environments. The training is episodic, with each episode terminating when the initial host either completes sending the payload to the exfiltration host or becomes isolated due to firewall actions. This setup creates a realistic scenario where the agent must balance the urgency of exfiltration with the need to remain undetected.

The architecture for both the actor and the critic in the RL model consists of a two-layer feed-forward neural network. The first layer of this network comprises 64 neurons, while the second layer contains 32 neurons. This configuration allows the model to process and learn from the complex patterns and strategies involved in successful data exfiltration within a networked environment.

7.5.3.2 Hyperparameters in Training

The training of the RL agent involves several key hyperparameters, carefully chosen to optimize the learning process. The critic learning rate and the actor learning rate are both set at 0.0003, ensuring a gradual and steady learning curve, which is crucial for the agent to effectively adapt to the varied scenarios it encounters in the network environment. The discount factor, a crucial element in RL algorithms, is set at 0.99, indicating a strong consideration for future rewards, encouraging the agent to plan for long-term outcomes rather than just immediate gains.

The horizon, or the number of steps in each episode before the gradients are computed, is set at 2048, providing the agent with a substantial window of experience to learn from within each episode. The minibatch size is chosen as 32, and the model is trained over 5 epochs, allowing for multiple passes over the training data for better generalization. The GAE parameter is set at 0.95, which helps in reducing the variance of the policy gradient estimates, and the clipping parameter at 0.2 is used to prevent too large updates to the policy, thus maintaining training stability. Finally, the entropy coefficient, a parameter crucial for encouraging exploration by the agent, is set at 0.02.

7.5.3.3 Training Episodes and Payload Targets

For practical application and evaluation, the RL agent is trained for 800 episodes on the first network and 1000 episodes on the second network. This difference in the number of episodes is reflective of the complexity and size differences between the two networks. The target payload for these exfiltration exercises is set at 10,000 MB, presenting a substantial challenge for the agent in terms of both volume and the need to evade detection while transferring this amount of data.

7.5.4 Experimental Results

The experimental results from the enhanced exfiltration path analysis study provide insightful observations about the performance and behavior of the RL agent across two distinct network environments.

7.5.4.1 Performance and Convergence

The training for both the first and the second network demonstrated stable progress with the RL policy converging efficiently. In the first network, which was smaller in scale, the RL agent's total rewards in an episode steadily increased to almost 12,000 over 800 episodes, indicating a significant improvement in the agent's ability to successfully complete the exfiltration task. Similarly, in the larger and more complex second network, the sum of rewards in an episode increased to just over 10,000 across 1000 episodes. Concurrently, the length of the episodes gradually decreased in both networks, suggesting that as training progressed, the RL agent became more efficient, completing the exfiltration task more swiftly and taking fewer random or unnecessary actions (Figures 7.7, 7.8 and 7.9).

7.5.4.2 Attack Path Analysis

In the first network, the RL agent's initial foothold was on host (8 2) in subnet 8, which was not directly connected to the Internet. From this position, the agent conducted subnet scans and discovered other hosts within the same and connected subnets, leading to the compromise of host (4 2) in subnet 4. This host became a key point for further exploitation to create an exfiltration path. However, the initially discovered path (8 2) -> (4 2) -> (2 0) was not fully aligned with the exfiltration

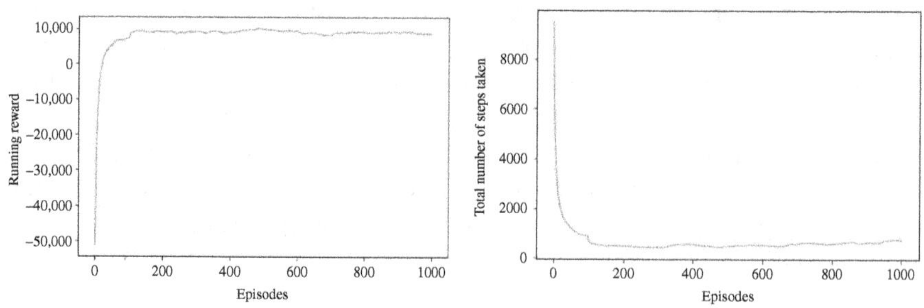

Figure 7.7 Enhanced exfiltration path training on first network [1].

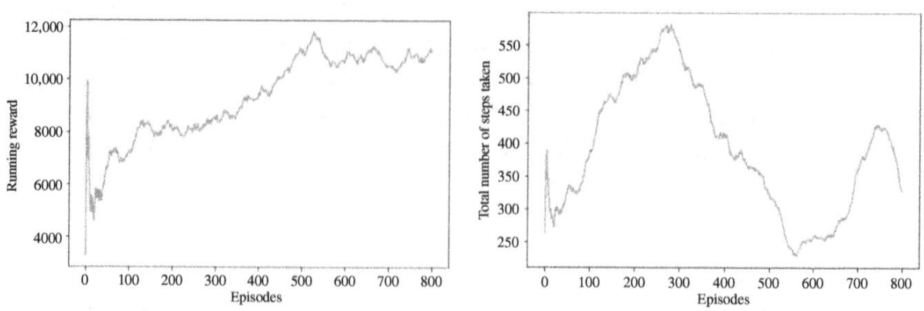

Figure 7.8 Enhanced exfiltration path training on second network [1].

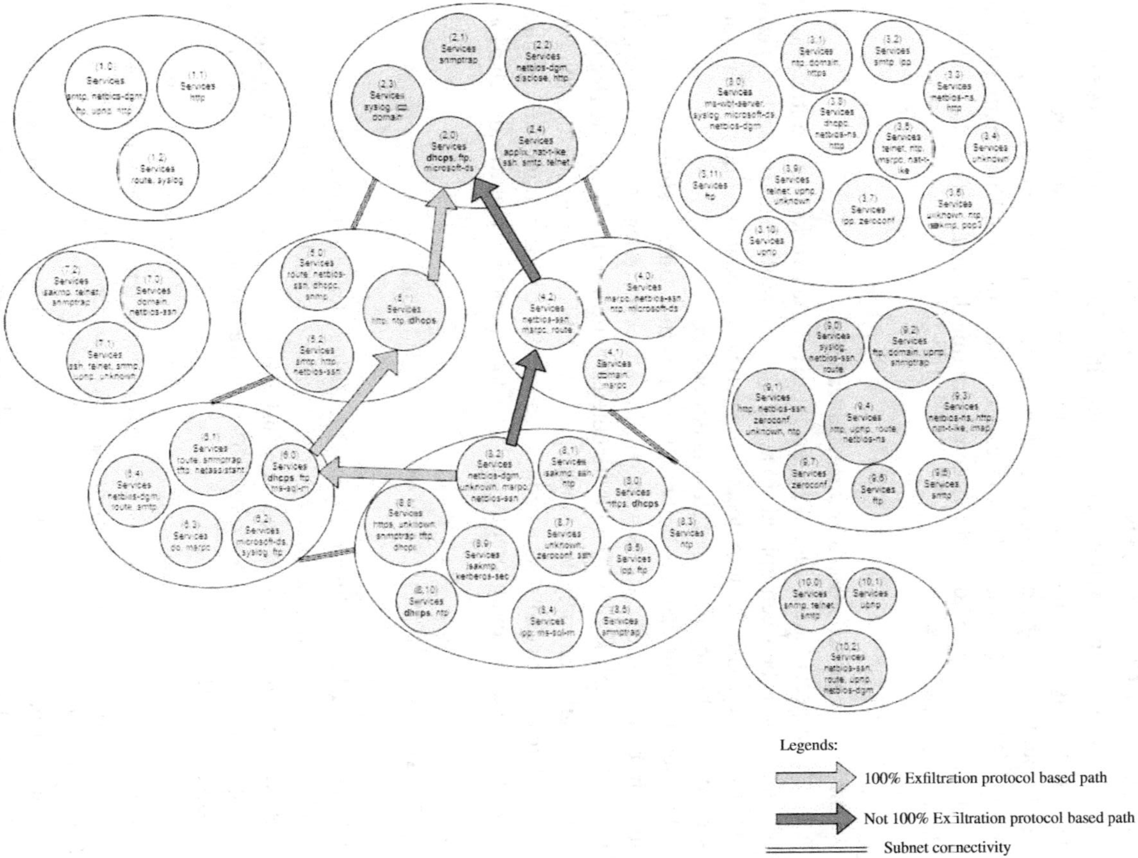

Figure 7.9 Enhanced exfiltration path analysis

protocol (DHCP) since host (4 2) did not run the DHCP service. In its search for a more optimal path, the agent then compromised host (6 0) in subnet 6 and eventually formed a more protocol-compliant path (8 2) -> (6 0) -> (5 1) -> (2 0), which utilized the DHCP service on both hosts (6 0) and (5 1).

In the second network, the agent started with a foothold over host (44 5) in subnet 44. Through various scans and exploits, the agent gained control over host (24 18) and eventually discovered and exploited the target host (5 10). The path formed (44 5) -> (24 18) -> (5 10) was a complete protocol-based path, utilizing the HTTPS service on host (24 18), demonstrating the scalability of the model in a larger network.

7.5.4.3 Strategic Actions and Protocol Utilization

A notable aspect of the RL agent's behavior in both networks was the strategic use of sleep actions between uploads to avoid unusual traffic patterns and triggering cyber defenses. This tactic reflects a sophisticated understanding of operational security and the importance of staying undetected.

Furthermore, the agent's ability to find paths that utilize a single network protocol in both networks aligns with real-life attacker strategies. By using existing network protocols known to network defenses, the agent reduced the risk of discovery by traffic anomaly detection algorithms. This approach of exfiltrating data using standard protocols, while considering traffic timing and volume, emulates tactics, techniques, and procedures (TTPs) documented in real-world scenarios.

References

1 Tyler Cody, Abdul Rahman, Christopher Redino, Lanxiao Huang, Ryan Clark, Akshay Kakkar, Deepak Kushwaha, Paul Park, Peter Beling, and Edward Bowen. Discovering exfiltration paths using reinforcement learning with attack graphs. In *2022 IEEE Conference on Dependable and Secure Computing (DSC)*, pages 1–8, 2022.

2 Rohit Gangupantulu, Tyler Cody, Abdul Rahman, Christopher Redino, Ryan Clark, and Paul Park. Crown jewels analysis using reinforcement learning with attack graphs. In *2021 IEEE Symposium Series on Computational Intelligence (SSCI)*, pages 1–6, 2021.

3 Lanxiao Huang, Tyler Cody, Abdul Rahman, Christopher Redino, Ryan Clark, Akshay Kakkar, Deepak Kushwaha, Paul Park, Peter Beling, and Edward Bowen. Exposing surveillance detection routes via reinforcement learning, attack graphs, and cyber terrain. In *2022 21st IEEE International Conference on Machine Learning and Applications (ICMLA)*, pages 1–8, 2022.

4 Xinming Ou, Sudhakar Govindavajhala, and Andrew W Appel. MulVAL: A logic-based network security analyzer. In *USENIX Security Symposium*, volume 8, pages 113–128. Baltimore, MD, USA, 2005.

5 Cheng Wang, Akshay Kakkar, Chris Redino, Abdul Rahman, S Ajinsyam, Ryan Clark, Daniel Radke, Tyler Cody, Lanxiao Huang, and Edward Bowen. Discovering command and control (C2) channels using reinforcement learning (Rl). In *Submitted IEEE SouthEastCon 2023*, 2023.

8

Using and Extending These Models

8.1 Supplementing Penetration Testing

As examined in Chapter 2, penetration testing requires completing a many-faceted and often manually intensive series of actions. Using reinforcement learning (RL) models in conjunction with a penetration test can improve the speed and accuracy of the testers. Some ways we can apply models to the testing will be readily apparent, but some limitations and benefits need consideration.

8.1.1 Vulnerability Discovery

A traditional penetration test report will list the vulnerabilities that were exploited, how they were discovered, and how they can be used by an actual attacker. At the most superficial level, the models described here will output the list of hosts and the vulnerabilities used to achieve the simulation goals. However, a straight export of this list has little value as a pen-testing report. The simulation should not be thought of as a substitute for a good penetration test but rather as an accelerator. Time is one of the scarcest resources during an operation. These models can explore a vastly greater number of options in a tiny fraction of the time and yet run offline in a network representation based on scans. The real-time simulation will still have to be performed by seasoned penetration testers, but the current model's output provides a way to speed up planning and initial network assessments.

While this simulation provides valuable information, it is not equivalent to the information provided by running the same exploits in real time. Many of the models described in previous chapters are based on a series of assumptions, and even in some far more mature future states of this framework. It is very unlikely that it will exist in a state without at least some assumptions. As described in Chapter 6, there is a sliding scale of realism in simulations, and though the methodologies

Reinforcement Learning for Cyber Operations: Applications of Artificial Intelligence for Penetration Testing, First Edition. Abdul Rahman, Christopher Redino, Dhruv Nandakumar, Tyler Cody, Sachin Shetty, and Dan Radke.

described in this text make great strides to advance this realism, it will always be imperfect as long as it is run offline. At the least, even with perfect information from scans, and perfect coverage, the scan information itself can't always be expected to be up to date, something can change between scanning and running the tests, so there is always implicit uncertainty.

The pen-testers should, therefore, attempt to follow some of the attack paths laid out in the simulation and see what works on the networks themselves. The human operators should not strictly explore the vast multitude of options the model can explore. Instead, the most appropriate application of these offline tests is for **planning** (Figure 8.1).

Planning is a critical phase in penetration testing, as it lays the foundation for a successful and effective assessment of an organization's security posture. In the comprehensive planning of a penetration test, several crucial elements must be addressed. Firstly, the test scope needs a clear definition, specifying the systems, networks, and applications within the testing boundaries. Simultaneously, rules of engagement, including the testing time frame, methods, and any restrictions or limitations, must be determined. Objective identification is equally vital, involving the clear articulation of test objectives, which may range from identifying vulnerabilities to testing specific security controls or simulating attack scenarios. Adequate resource allocation, spanning personnel, tools, and equipment, is essential to ensure the testing team possesses the requisite skills for an effective assessment. Legal and ethical considerations play a pivotal role, demanding the penetration test be conducted within established boundaries. This includes obtaining proper authorization from the organization being tested and adhering to all relevant legal and regulatory requirements.

A comprehensive risk assessment is conducted to identify potential risks and understand their impact on the organization's systems and operations. Information gathering involves strategic reconnaissance to acquire details about the target environment, encompassing IP addresses, domain names, and network infrastructure. Defining the testing methodology is a crucial aspect, involving the selection of appropriate tools and techniques for identifying vulnerabilities and their exploitation. Effective communication planning is established, outlining how information will be shared between the testing team and the organization, including status

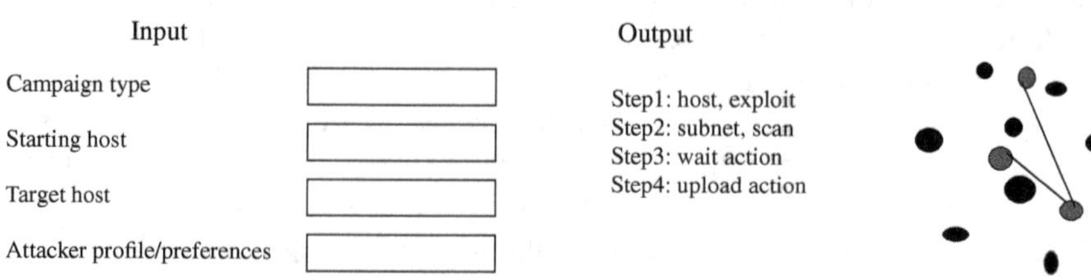

Figure 8.1 UI Mockup of penetration testing planning application.

updates, incident reporting, and post-test debriefings. Contingency planning is developed to address unexpected situations during the penetration test, ensuring a swift and effective response to challenges. Lastly, documentation requirements are defined, encompassing a detailed test plan, documentation of the testing process, and a comprehensive report of findings and recommendations.

The ability to generate attack paths, even imperfectly and with errors, can greatly help accelerate this planning phase. By the time the penetration testing team gets the model's output, the model has run thru 1000s of iterations by trial and error.

By thoroughly planning a penetration test with the assistance of models, organizations can ensure that the assessment is conducted in a systematic and controlled manner, maximizing the effectiveness of the testing process and minimizing potential risks and disruptions to business operations.

8.1.2 Path Analysis

Understanding what vulnerabilities are present in a network is key to building a good defensive posture, but it is far from the end of the storing, patching, or even monitoring every vulnerability after it has been identified is not always possible. Managing security for real networks always involves compromise as time and resources are always finite. Sometimes, there are constraints to use legacy systems or a limited budget for sensor placement, and remediation of vulnerabilities becomes more difficult. In situations such as these, it helps to understand the larger **context** of the vulnerabilities, and how they impact the wider system. Understanding what attack paths the vulnerabilities lie along provides valuable context for how severe a threat is actually present. If you have lost the key to a lock in your home, this information by itself is not very actionable. Knowing the key was for the front door, or, say, the master bedroom, has very different implications. Understanding the attack path can give insights similar to this. Perhaps a chain of vulnerabilities is required to gain access to a certain host, and improving security along any point on that path may be enough to mitigate the risk. Understanding a single path may not be enough either. If multiple paths exist and they cross through a common host, exploiting a common vulnerability, this can also inform how to prioritize remediation with limited resources.

Path analysis goes beyond identifying individual vulnerabilities. It can reveal the intent of the attacker, helping organizations prioritize defense against specific threats. For instance, a business may be able to handle unauthorized access or leaked customer data, but a brief service outage could be catastrophic. The level of accepted risk varies based on the business, the network, and the context. Path analysis provides the necessary insights to make informed decisions in such scenarios.

Path analysis isn't unique to this methodology, and any good penetration test should involve some path analysis. However, RL allows path analysis that is informed by context, scalable, and can be prioritized by smarter risk scoring.

Path analysis in penetration testing involves the systematic exploration of potential attack paths that an adversary might take to compromise a target system or network. The goal is to identify and analyze the various steps or stages that an attacker could follow to achieve a specific objective. These objectives potentially include gaining unauthorized access, extracting sensitive information, or disrupting services. Path analysis is a key component of the overall penetration testing process and helps security professionals understand and mitigate potential security risks.

The path analysis in penetration testing unfolds through several key phases. Beginning with reconnaissance, the penetration tester collects information about the target system, network, and organization, identifying potential entry points, IP addresses, domain names, and other pertinent details. Subsequently, the enumeration phase actively discovers and identifies resources within the target environment, including active hosts, services, and open ports. Moving to vulnerability analysis, the tester assesses weaknesses using scanning tools, manual testing, and other techniques to uncover potential entry points. The exploitation phase follows, where the penetration tester attempts to exploit identified vulnerabilities, employing exploit scripts, malware, or other attack techniques to gain unauthorized access or compromise the target system. The lateral movement comes into play after establishing an initial foothold, exploring potential pathways an attacker might take within the network to escalate privileges or compromise additional systems. Seeking to maintain access over time, the persistence phase identifies methods an attacker might use, such as creating back doors or planting malware. For scenarios involving data theft, path analysis examines potential paths for data exfiltration, including data sources, transmission methods, and exit points. Covering tracks is the final step, where an attacker may erase evidence of their activities using techniques like modifying log files or deleting artifacts. Throughout this process, documentation is paramount, with the penetration tester documenting each step, including attack paths, vulnerabilities, and successful exploitation. This documentation forms the basis for creating a comprehensive penetration testing report. Path analysis provides valuable insights into the security posture of an organization by simulating potential attack scenarios. It helps organizations understand their vulnerabilities and weaknesses, allowing them to prioritize and implement security measures to better defend against real-world threats.

There are currently existing tools to assist in path analysis. BloodHound [1] performs path analysis in active directory (AD) environments by leveraging graph theory and analyzing the relationships and permissions within the AD structure. The tool collects and processes information about users, groups, computers, permissions, and other AD objects, allowing security professionals to visualize and understand potential attack paths, privilege escalation opportunities, and lateral movement possibilities. Here's a step-by-step explanation of how BloodHound conducts path analysis:

Data collection in BloodHound involves querying various aspects of the AD environment using methods such as LDAP queries, powershell, and SMB queries. This encompasses gathering information about users, groups, computers, permissions, and group memberships. Following data collection, BloodHound processes the information to construct a graph representation of the AD environment. The graph comprises nodes representing AD objects like users, groups, and computers, and edges representing relationships or permissions between them. BloodHound utilizes a graph database, commonly Neo4j, to efficiently store and manage the collected data, facilitating effective querying and relationship analysis. The tool identifies and analyzes relationships within the AD structure, including group memberships, user-to-computer relationships, user-to-group relationships, and other permissions and trust relationships. BloodHound further calculates potential attack paths by analyzing the graph, identifying sequences of relationships and permissions exploitable by attackers, such as paths leading to privilege escalation or lateral movement. The graphical interface provided by BloodHound visually represents the AD graph and identifies attack paths, aiding security professionals in understanding complex relationships and potential security risks. The tool incorporates a query language for users to create custom queries tailored to specific aspects of the AD environment. BloodHound generates reports highlighting identified security risks and potential attack paths, offering remediation recommendations to assist security teams in addressing and mitigating vulnerabilities. Furthermore, BloodHound can be employed for continuous monitoring, enabling security professionals to track changes in the AD environment and detect new attack paths as the network evolves.

By combining the power of graph theory with data visualization and analysis, BloodHound provides a comprehensive view of an organization's AD structure, facilitating effective path analysis for penetration testing and security assessments. It helps organizations proactively identify and address potential security risks within their AD environments. While this is powerful and useful, this is driven what amounts to rules, what's allowed, and has no information regarding motivation, stealth, or prioritization. BloodHound provides several features to help prioritize and assess the risk associated with attack paths identified within an AD environment. While the tool itself doesn't inherently assign priorities, it offers insights and data that can be used by security professionals to make informed decisions. Here are some factors and features that contribute to prioritizing attack paths in BloodHound:

- **Shortest Paths:** BloodHound identifies the shortest paths to specific objectives, such as domain admin privileges. These paths represent the most direct routes an attacker could take to achieve a particular goal. Shorter paths are generally considered more critical and may be prioritized for investigation and remediation.

- **Risk Scoring:** Some versions of BloodHound include risk scoring features, where the tool assigns scores to different paths based on the severity of the associated security risks. This can help prioritize paths with higher risk scores for immediate attention.
- **High-Value Targets:** BloodHound allows users to identify and focus on high-value targets within the AD environment. For example, paths leading to critical systems or sensitive data may be considered higher priority than paths leading to less critical assets.
- **Graph Visualization:** BloodHound's graphical interface provides a visual representation of the AD graph and attack paths. Security professionals can visually inspect the graph to identify complex relationships and prioritize paths that involve critical AD objects or sensitive permissions.
- **Custom Queries:** Users can create custom queries in BloodHound to filter and prioritize paths based on specific criteria. This flexibility allows security professionals to focus on paths that align with their organization's security priorities.
- **Impact Assessment:** BloodHound assists in assessing the potential impact of attack paths. Security teams can prioritize remediation efforts based on the potential business impact by understanding the implications of a successful attack on specific paths.
- **Remediation Recommendations:** BloodHound provides recommendations for remediation, helping security professionals address and mitigate the identified vulnerabilities associated with attack paths. Remediation efforts can be prioritized based on the severity and criticality of the issues.
- **Continuous Monitoring:** BloodHound supports continuous monitoring of the AD environment. Security teams can use this feature to track changes over time, identify new attack paths, and adjust priorities based on the evolving security landscape.

It is important to note that the prioritization of attack paths should be contextualized within the specific goals and risk tolerance of the organization. Security professionals should consider factors such as business impact, critical assets, and regulatory compliance when prioritizing and addressing identified vulnerabilities. BloodHound's features provide valuable information that, when combined with expert analysis, can guide organizations in effectively managing and mitigating security risks. Contextualization and risk scoring beyond the atomic vulnerabilities is where artificial intelligence (AI)-driven methods, like those described in this text, can provide a strong advantage.

8.2 Risk Scoring

8.2.1 Current State

A risk score for cyber is often described as being analogous to a credit score (citations, several), it takes a variety of data into account and produces a scalar measurement of the security of a system, process or asset, providing a simplified summary of the state of things, but meant to be informative. One common framework used in risk analysis is FAIR. FAIR, which stands for factor analysis of information risk, is a quantitative risk analysis framework designed to provide a structured and financially oriented approach to understanding and measuring information security risk. Developed by the open group, FAIR focuses on expressing risk in monetary terms, allowing organizations to make more informed decisions about risk management and resource allocation.

Key components of FAIR Risk Scoring include loss event frequency (LEF), representing the probable frequency of a specific loss event occurring over a given time period, measured in events per year and based on historical data, threat intelligence, and expert judgment. Threat event frequency (TEF) measures the probable frequency of threat events leading to the loss event, considering the threat landscape and potential threat actors. Vulnerabilities, both primary and secondary, contribute to the risk scenario, and control strength (CS) assesses the effectiveness of existing controls in percentage terms. Loss magnitude (LM) represents the potential financial impact of a loss event, encompassing direct, and indirect costs. In the risk analysis steps, scenario definition involves identifying and defining the risk scenario components. Data collection gathers relevant information, and calculations determine LEF, TEF, vulnerabilities, and control effectiveness. Risk is then calculated by combining these factors to determine the annualized loss exposure (ALE). Sensitivity analysis assesses the impact of variable changes on overall risk. The benefits of FAIR risk scoring include expressing risk in monetary terms for better communication, prioritizing risks based on financial impact, supporting risk treatment decisions, and facilitating scenario comparisons to aid decision-making and resource allocation.

FAIR provides a systematic and repeatable method for organizations to assess and manage their information security risks in a way that aligns with business objectives and financial considerations. It is widely used in industries where financial quantification of risk is crucial for decision-making, such as finance, healthcare, and critical infrastructure. There are, however, many ways to assign a risk score. Most often each vendor of a cybersecurity tool or stack of tools has their own method for assigning risk scores (citations, needed), and it can be difficult to compare between methodologies.

Other methods for risk scoring often involve a risk matrix. A risk matrix is a visual representation of risk that helps organizations assess and prioritize risks based on their likelihood and impact. In the context of cybersecurity, a risk matrix is a valuable tool for making informed decisions about which risks to address first and how to allocate resources effectively.

A risk matrix comprises two key components: Likelihood, representing the probability or frequency of a risk event, categorized as rare, unlikely, possible, likely, or almost certain; and impact, measuring the potential harm, including financial losses, operational disruptions, and reputation damage, categorized as low, medium, high, or critical. To create a risk matrix, clear definitions for each level of likelihood and impact are established. Likelihood levels, for instance, could range from rare (1) to almost certain (5), and impact levels from low (1) to critical (4). A matrix grid is then created with likelihood on one axis and impact on the other, forming a grid where each cell represents a combination of likelihood and impact. Risk levels or scores are assigned to each cell based on this combination. Risks are plotted on the matrix by assessing their likelihood and impact using analysis, expert judgment, or available data. In interpreting the matrix, high-risk areas are those falling into cells with both high likelihood and high impact, demanding immediate attention and mitigation efforts. Medium-risk areas involve risks in cells with either medium likelihood or medium impact, monitored closely with mitigation based on available resources. Low-risk areas encompass risks with low likelihood and low impact, not requiring immediate attention but still necessitating monitoring to ensure they remain at acceptable levels.

A shortcoming of these methodologies is that in summarizing, there is necessarily a loss of information. We cannot summarize without throwing something out. To provide quantitative repeatable results, these existing methods often lack what has been emphasized throughout this text: **context**.

8.2.2 Future State

To understand the shortcomings of current risk scoring methodologies, it helps to again turn to the credit score analogy. Someone with a poor credit score might find it an unfair description of their financial fitness. This unfortunate individual might argue that given their circumstances and understanding their story, one would understand that is safe to give them a loan or a good interest rate. A lender, however, would rather err on the side of caution, and these scores are useful tools for financial institutions, protecting them from risk, even though they may oversimplify the story of the individuals who get scored. The context of the individual being scored is washed away in the summary, but this is fine for the financial institution because they assume the risk. For a cybersecurity risk score though, there could be multiple types of consumers. A business may want to assess the security of a third party with whom they have dealings, perhaps with whom they share data. In this case, the business is analogous to the financial institution doing the lending, and it relies on the score to judge if it should proceed with an external entity. However, the more common case is that the consumer of the risk score is the organization being scored because they want to understand the risk of how they currently operate. Grading oneself is very different than grading others. If we were to give ourselves a credit score, we would take care to understand the context.

To begin down the path of improving cyber risk scores, it helps to take a step back and ponder the question "the risk of **what**." Risk is more accurately a vector, not a scalar; one can be very secure against certain kinds of threats and not others, such as having strong resilience to downtime but still having poorly guarded PII.

Understanding the **distribution** of simulations, the variety of attack paths that are viable and likely in a network is key to operational readiness. Methods leveraging attack graphs have a utility that is actually **limited** by how wide its scope is. Telling a defender **all** possible attack options in a vast space is not easily actionable, but conversely, having just a single attack path, even a realistic one, only tells a small part of the story, a single example of how things could go bad. What's likely more valuable is the **range** of things most likely to go wrong. It may be possible to apply some form of Bayesian learning on top of these simulations to understand some limits on these ranges of outcomes. Understanding vulnerabilities on hosts in the context of **multiple** paths that are **likely** can allow aggregate scores that are still informed by context.

8.3 Further Modeling

The narrative of this text centers on models that perform threat simulation but there are great number of other cybersecurity models that have synergies with this framework, some of these exist in some form today, and some of these purely future state.

8.3.1 Simulation and Threat Detection

Up to this point, modeling described in this text has focused on simulation and preventive actions that can be informed by that simulation. All of this is what happens before an attack, but machine learning (ML) can also greatly enhance the defensive capabilities for what happens during an attack, as it is ongoing. Threat detection is a field of research unto itself, and it is quickly evolving for reasons already stated in previous sections: cybersecurity is an arms race. Threats evolve over time, and the responses evolve as well. This back and forth isn't simply because old threats are completely degfeated or old defenses are made completely obsolete. Repeating ourselves again, cyber warfare is a game that is played very inefficiently. A new effective attack doesn't have to be brilliant or inventive, it simply has to be something the defense isn't looking for. The MITRE ATT&CK framework provides a powerful framework for classifying and understanding attack patterns, but all serious cybersecurity professionals understand its limitations. It is a model of how attacks can happen, and though some models are useful (and certainly this one is), no model is *true*. When the Russian invasion of Ukraine began, news of ransomware attacks by Russian operators circulated in the media, but a great number of the attacks from Russian APTs were those that exist outside of the Mitre ATT&CK framework. In recent years, there has been a shift in focus toward endpoint detection, and away from flow-based security, due to cost, size of data for growing networks, and more distributed networks as businesses increasingly move more and more operations to the cloud. This seems like a good strategic move, but it will also, inevitably lead to attacks designed to be invisible to the current tools, this is the game. In this game, where the target is always moving, signature-based approaches are always playing catch up for both offense and defense. The anti-virus (AV) running on your device that you are reading this on is mostly likely checking against a vast number of signatures for previously seen threats. This does have a lot of power and will not go away anytime soon, but it also has clear limitations in this ever-evolving fight. A shift toward behavioral-based detection, whether from rules in an intrusion detection systems (IDS), a simple anomaly detector, or a sophisticated deep learning model, can protect against what signature-based approaches fail to see. In the analysis of behavior, it is here that our simulations from RL intersect with threat detection, and this is an area that is ripe for exploration in research, with very little foundation currently at the time of writing.

8.3.2 Ransomware Detection

As explored in [15, 17], ransomware attacks commonly occur in several stages. The first stage involves the initial disbursement of malware to a targeted device using exploit kits, phishing emails, or several other intrusion methods. The next phase is when ransomware is installed on the machine, and infecting a target's network begins. From this point, the next phase is staging the ransomware, potentially through command and control (C2) communication and other methods. This phase aims to ensure that the malicious code can survive reboots. The next step in a ransomware attack involves identifying and encrypting sensitive data. This final step also delineates between multiple types of ransomware attacks, as malicious actors can take various actions after the data has been identified and accessed. For example, in a Doxware attack, hackers may threaten to publicize users' private data. In a traditional crypto-ransomware attack, data will be encrypted without necessarily affecting the user's systems.

This general structure can still allow for multiple different types of attacks to occur during a ransomware campaign. Ransomware taxonomies often show multiple different attack types within the infection stage, for example, and the same can be said of C2 communication. [11, 17]. Given these different attack patterns, ransomware attacks become more difficult to detect before the final encryption phase. For example, individual detections of attacks may indicate network infection based on suspicious activity on hosts. Still, they likely will not possess the necessary information to indicate whether that infection is part of a ransomware attack.

A ransomware detection model could synergistically synergize with a ransomware **simulator** powered by RL. Some possible benefits include:

- Detection of longer campaigns with partial signals, contextualized threats, and behavioral signals.
- Improving simulation by modeling more partial information based on the performance of threat detection models in the wild.
- Using simulation in conjunction with classification models to understand a threat as it happens, to know where the attacker will likely move next so that effective triage is possible.
- Through synthetic training data generated by simulation, there is the possibility of building a detector capable of attribution.

8.3.3 Engineering New Exploits

The primary use of the overarching RL framework described in the bulk of this text seeks to uncover vulnerabilities in networks, but these are emergent vulnerabilities that come about from the alignment of more atomic exploits. The model described thus far may be creative in orchestrating the available exploits, but it cannot "invent" new exploits, and it is restricted to the options explicitly listed in its action space. A model supporting vulnerability discovery at the atomic level should be possible. Red teaming professionals and researchers always uncover new vulnerabilities, and they do not just emerge from the ether. As new software rolls out or old software is updated, the right professionals with the right skills in the tradecraft can bend tools to be used in unintended ways. This tradecraft involves experimentation and creative thinking, but playbooks, methods, and encoding to this art can be translated to a set of actions, states, and rewards for yet another RL model to play with. Such a model would automate penetration testing on a more granular level, discovering the vulnerabilities of individual hosts, individual packages, or versions of packages, a layer deeper than the aggregate network vulnerabilities that have dominated much of the conversation here so far. The immediate synergy of such a vulnerability discovery model on top of an attack simulation model is obvious: a dynamic way to extend the action space and give our exploitation agent more options and more tools. Such a model could also feed into the same threat detection models discussed earlier. In rules-based approaches, which rely on specific signatures and IOCs, such a vulnerability discovery model could provide additional rules to throw on top of the pile or enrich currently deployed ones. However, as alluded to in previous sections, the inflexibility of rules approaches is something that advanced cybersecurity will continue to move away from as the arms race continues to accelerate. In a behavior-driven threat detection approach, a vulnerability discovery model can still provide valuable input by informing the features of such a model.

8.3.4 Extension to LLMs

The recent spotlight given to large language models (LLM) raises many questions about their impact on the attack surface, as organizations are still learning to integrate these tools into existing systems or replace said systems entirely. While the specific types of attacks may change and the way such software interacts with them are different, in principle, the way an AI would interact with this is not vastly different. The models described in this section for threat simulation, threat detection, and vulnerability discovery can all have language model-based flavors to them and be integrated with their vanilla counterparts. A set of actions to compromise a system leveraging an LLM using known exploits of that particular LLM version would, in principle, operate no differently from what's been described in this text. The discovery of new vulnerabilities in LLMs sees some shift in that the toggles to play with this particular piece of software are now a series of text prompts, but in principle, the experimentation with different configurations to use the software in unintended ways carries much the same thematic character.

8.3.5 Asset Discovery and Classification

An assumption that runs through this text is that the high-value target in an operation is known at outset. In a crown jewels analysis simulation, the location of the crown jewels is given as input. In an exfiltration simulation, the host, which acts as the starting point to exfil from, is assumed. Depending on the attacker's prior intelligence, these may be reasonable assumptions to make, but other times they are not. Often an attacker will gain access to a network and simply know the information they want is somewhere, not a specific table, machine or subnet. The structuring agent described in earlier chapters explores and gathers information about the network it is attacking, but the information it gathers is just at one layer, telling it what actions are possible at the next step, but not telling it about the identity, assets of specific hosts as it bounces through a network. A real attacker must often figure these things out on the fly as they usually have limited information about what is on the network they are traversing; in fact, most network owners have incomplete information about what is on their networks themselves. A lot can be inferred from the same scan information gained leveraged in the models discussed in the rest of this text. If a server is running database software, there is a chance that it is used to store critical organizational data like client PII or information about the personnel of the organization. Further scans, on top of learning about the services and software versions on the host, can be used to determine the role of that host and the nature of its assets. The additional scans, of course, are another opportunity for an attacker to be discovered if they are accessing information in an unusual way for the network. These scans alone are also likely to be insufficient in making a clear determination. A file will probably not be named "PII in here." Another model, on top of an RL framework for threat simulation, can be used to classify assets as they are discovered. Such a model can be trained in a supervised fashion based on how different assets appear on different networks and what context clues they leave based on the services on the host, the identified security to access the assets, and actual reads of data or metadata about these assets (Figure 8.2).

8.3.6 Attribution

"Attribution matters because it allows you to think about how you can best strategize and predict future attacks." [10].

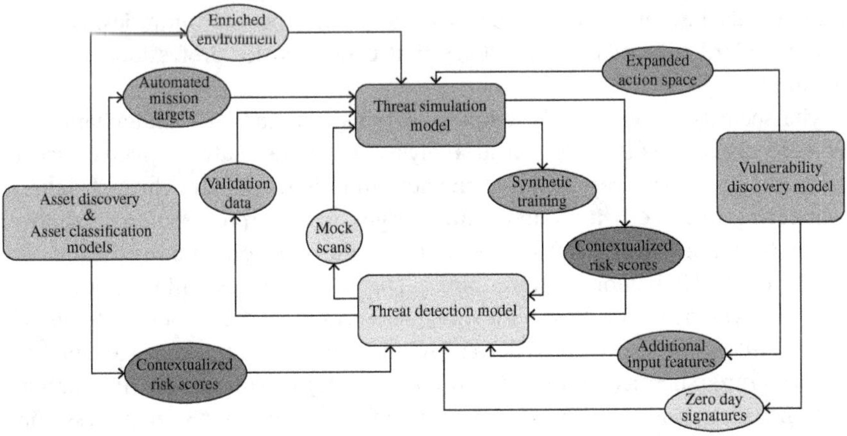

Figure 8.2 An example diagram of how multiple cyber defense models could interact, as compartmentalized parts of understanding how an attack actually happens. This looks advanced and complicated compared to the current state today, but even this aspirational image is still analogous to having separate brains for walking and chewing gum.

Understanding who is conducting an attack isn't just about blame (though this can be important, too, politically). Understanding who is targeting you can give insight into the attacker's motivations, and as we have already discussed in Chapters 6 and 7, motivation greatly informs how an attack is conducted. Knowing the enemy is helpful before, during, and after an attack. A complete understanding of what occurred during an attack may not happen until after the malicious behavior is detected. Proper attribution can help to accelerate this process. During an attack, knowing the adversary helps decide which defensive countermeasures to undertake as a detected threat is actively moving through your network. Once you know a particular group has targeted you, defense against future attacks can be planned much more intentionally.

Attribution is no trivial matter, however. It is considered to be among the two holy grails in cybersecurity; the other being attacker intention. Different groups use different tools and strategies, but these rarely have distinct smoking gun signals, and high confidence attribution usually comes from a holistic approach, looking at all the possible data available and correlating it with past attacks from different groups. Here is another area where simulation could significantly contribute, especially in conjunction with additional models. Understanding ahead of time how different APTs may operate within **your** network through simulation can make it easier to do real-time and give attribution.

Attribution in the context of cybersecurity refers to identifying and assigning responsibility to individuals, groups, or nation-states behind cyberattacks. Attribution is a complex and challenging task that often requires a combination of technical analysis, intelligence gathering, and collaboration among cybersecurity professionals, law enforcement agencies, and intelligence communities. Here are some standard methods and practices that cybersecurity professionals use for attribution:

Cybersecurity professionals employ various methods to understand and attribute cyber threats in technical analysis. Malware analysis involves scrutinizing malicious software's code, behavior, and characteristics through reverse engineering to discern its functionality, origin, and purpose. Forensic analysis focuses on digital artifacts left after an attack, aiding in reconstructing the timeline and identifying indicators of compromise. Digital signatures, unique patterns, or fingerprints associated with specific threat actors or malware are sought for identification. Infrastructure analysis delves into identifying and tracking the attackers' infrastructure, such as command and control (C2) servers, using IP addresses, domain registrations, and hosting providers for attribution. Cyber professionals unravel proxy chains and anonymization techniques attackers use to conceal their location. Tactics, techniques, and procedures (TTPs) analysis involves studying attack patterns and revealing specific threat actors or hacking group associations by analyzing tools, processes, and procedures. Comparing current attacks with historical incidents helps identify commonalities and linkages to known threat actors. Open Source Intelligence (OSINT) involves leveraging publicly available information from the internet and social media to gather intelligence on potential threat actors. Threat intelligence feeds provide real-time information on known threat actors and their tactics. Collaboration and information sharing, both within the industry and with government agencies, enhance attribution capabilities through collective intelligence. Finally, behavioral analysis, focusing on threat actors' motivations, targets, and techniques, provides indirect but valuable insights into their identity, contributing to a comprehensive profile.

It is essential to note that attribution in cybersecurity is inherently challenging, and false flags, deception techniques, and proxy infrastructure can complicate the process. As a result, attribution is not always conclusive, and cyber professionals often use terms like "high confidence" rather than absolute certainty. Additionally, public attribution statements are typically made by government agencies, security firms, or international organizations rather than individual cybersecurity professionals.

Attribution in cybersecurity is a complex and challenging task, and while there is no definitive method to attribute attacks with absolute certainty, several models and tools are used to assist in the process. These models and tools leverage various techniques, including technical analysis, threat intelligence, and collaboration. The diamond model of intrusion analysis offers a comprehensive framework for dissecting cyber threats, emphasizing four crucial elements: adversary, infrastructure, capability, and victim [21]. This model assists analysts in structuring and visualizing information, supporting attribution efforts. The cyber kill chain, devised by Lockheed Martin, delineates the stages of a cyberattack, aiding in pattern identification and attribution through the examination of TTPs used at each stage. The MITRE ATT&CK Framework provides a detailed matrix of adversarial actions across the attack lifecycle, facilitating understanding, and attribution based on observed TTPs [22]. In the realm of attribution tools, threat intelligence platforms like ThreatConnect, ThreatStream, and recorded future aggregate and analyze threat intelligence data, aiding in pattern recognition and linking threat actors. Maltego, a data visualization tool, assists in exploring relationships between entities for a comprehensive view of threat actor infrastructure. Platforms like CrowdStrike Falcon Intelligence, Recorded Future, and FireEye contribute to attribution by providing real-time threat information and understanding the tactics of threat actors. Shodan serves as a search engine for internet-connected devices, helping identify exposed infrastructure and potential attack targets. VirusTotal enables analysts to submit files or URLs for analysis, providing insights into associations with known malware or threats. OSINT tools, including Maltego, SpiderFoot, and theHarvester, gather information from public sources, enhancing the understanding of threat actor context and motivations. Together, these tools and frameworks empower cybersecurity professionals in the complex task of attributing cyber threats.

It is important to note that attribution is challenging, and these tools and models are used collectively to build a comprehensive understanding of cyber threats. False flags, deception techniques, and the use of proxy infrastructure by adversaries can complicate the attribution process. Analysts often combine technical analysis with intelligence gathering, collaboration, and contextual understanding to provide informed attribution assessments. Additionally, attribution is often a responsibility shared among government agencies, private cybersecurity firms, and international organizations.

The AI-driven framework proposed here can help with attribution. Training data for labeled attacks is scarce enough, but it is especially scarce to be labeled as belonging to a specific threat actor. If common strategies of a given threat actor are understood even on a high level, then an RL model similar to those described in this text could generate synthetic data driven by rewards engineered based on the ATP's preference. A deep neural network classifier can compare live sensor data (flows/endpoint) to these simulations with this synthetic data. A metric learning classifier that embeds a path (or part of a path) to a point in learned space can then see if a piece of the attack path looks like the paths we have generated and then can determine the closest ATP and, from a distance, assign a confidence score (Figures 8.3 and 8.4).

Let us unpack this notion a little further. Attributing detected behavior in a given network topology consists of two interdependent pieces: the ability to detect malicious, often nuanced, behavior from a threat actor in a network, and the ability to

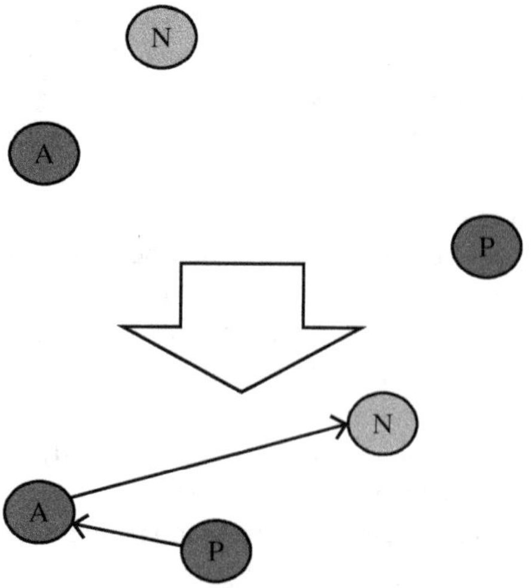

Figure 8.3 An example of metric learning, specifically triplet loss, where examples belong to similar classes (in blue) are pulled together in the latent space as the model learns, while simultaneously pushing examples from different classes (in red here) are pushed farther away. The technique is named for computing the loss based on sets of three examples: an anchor "A" for reference, a positive "P" of the same class, and a negative "N" of some other class.

correlate that behavior to predetermined attack paths based on RL simulation. We will explore both of these pieces independently.

8.3.6.1 Detecting Malicious Behavior

Although beyond the scope of this work, we will briefly explore the several methods through which behavior from threat actors can be detected in a network. Broadly speaking, these methods fall into one of three categories: tool-based, rules-based, or AI-based methods.

Tools-based capabilities, as the name suggests, rely on the out-of-the-box detection capabilities present in commercially available off-the-shelf (COTS) security tools. Some examples of these could include endpoint based threat detection tools like Crowdstrike Falcon or SentinelOne Singularity Endpoint, or network detection and response (NDR) tools such as Darktract Detect or Vectra AI. Each of these tools has one or many MITRE ATT&CK TTPs and contains a multitude of proprietary methodologies to detect threats that include custom logic, integrations with OSINT or proprietary threat intelligence feeds, or in-house AI techniques for detecting threats. These capabilities are offered out of the box, but at the cost of limited customizability or visibility into the algorithms' inner workings.

Rules-based capabilities offer an additional layer of customizability to tools-based techniques by offering defenders the ability to write logic to detect certain patterns of activity in cyber telemetry. Rules-based capabilities are often rudimentary and are generally limited to matching patterns with regular expressions or looking for certain kinds of events in given time windows. Consequently, rules are very effective at identifying known signatures of threats such as malicious file hashes or IP addresses but can be brittle or prone to false positives when attempting to perform behavioral analyses of users or entities (UEBA). It should also be noted that rules-based capabilities are often available with COTS tools or other open-source security tooling, and several open-source accelerators, such as starter kits are also widely available. However, readily available rules can require significant tuning before proving efficacious in production environments.

Finally, custom AI-based methodologies refer to the paradigm of deploying and maintaining custom AI models for detecting cyber threats on a given network that work in tandem with, or on top of, COTS tools, security information and event management (SIEM) systems, or security orchestration, automation, and response (SOAR) systems. There has been abundant work in the literature on effective methods for AI-based threat detection [2, 4, 5, 9, 18], and such methods are often viable in detecting behavioral threats or anomalies that require a more nuanced understanding of threat vectors that are simply too complex for traditional rules-based approaches. Deploying and maintaining effective AI-based methods are often research and resource-intensive tasks that require operator feedback during deployment stages to hone in on their efficacy. However, if done correctly, AI-based approaches can offer robust, actionable detections that drastically improve the security posture of defending organizations.

Clearly, each of the above approaches has its benefits and drawbacks, but maintaining a strong security posture as a defender requires the curation and maintenance of all of these tools working together to maximize coverage and effectiveness.

8.3.6.2 Attributing with Attack Paths

In order to correlate disparate detections from the aforementioned tools, all or any combination of which may be present in enterprise environments, we must first define a standard method for combining detections into a single data schema. One approach to do so could be to rely on COTS tools such as Google's Chronicle SIEM or other extended SOAR (XSOAR) tools to combine disparate logs. Alternatively, an ML-based approach such as that proposed by Murli et al. [12] could combine COTS and AI-based detections into "alert graphs."

Once all alerts are unified into a single representation across tools, those events could then be passed through AI-based methods and models that utilize metric learning to separate attacks by different adversaries in a latent space, some of which have been performed in several recent works [12–14]. This would, in essence, provide us with a well-separated latent space in which attacks, at their various stages, by various adversaries, are grouped by similarity but are wholly distinct from each other. Parts of attack graphs or paths could then be embedded in the same latent space while running the model in production, and a computation of closest neighbors (or cluster classification) could help attribute said attack path to the behavior of an adversary seen during training.

The RL methods proposed in this work could further add training data to the "alert graphs" mentioned above by simulating several possible, realistic, combinations of attack paths and techniques for any adversary. This would allow for the metric learning model to learn a robust, generalizable latent space which can be used for zero-shot attribution of novel or partially reused campaigns in real networks (Figure 8.4).

8.3.7 Defensive Modeling

Penetration testing and red teaming, in general, are meant to improve the defensive posture of a network by understanding and addressing its weaknesses by reacting to the information gained from the mock malicious activity. Up to this point, we've focused on attack simulation but have said very little about the reaction part, about what defensive countermeasures can be made in response to these attacks. There has been a little bit of defensive modeling mentioned earlier, in some of the hard-coded reactions of the network to attacks, such as firewall

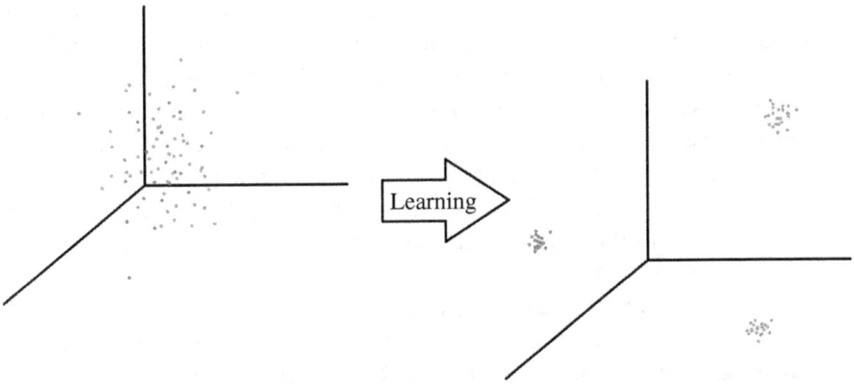

Figure 8.4 The aggregate result of metric learning is a latent space with meaningful arrangement and distances (metrics) between examples such that the space encodes a conceptual understanding. The colored dots representing examples belonging to different classes are sorted and separated out as the underlying conceptual idea of the different classes becomes more crisp and defined in the model.

updates isolating hosts in response to activity as in [20]. Such hard-coded behavior adds some realism to these offline simulations by emulating the dynamic nature of these conflicts. To enhance this realism further, instead of hard coding such defenses, they, like the offense can carry out its actions as dictated by AI. Besides further enhancing the realism of simulation by further capturing what these attacks would look like in practice, there is also an opportunity here to extract defensive recommendations from the network, where to place what sensors, what tools, and what rule sets to apply.

8.3.8 AI vs. AI

The need for **realism** in simulations has been something emphasized repeatedly in this text, and from the authors' perspective, this has meant making simulated attacks more like those of human attackers. But as stated previously, this is an arms race, ever-evolving, and with the rise of AI-driven defenses and attack simulations, so too will true AI attacks emerge. Like many topics here, this is still in its infancy. If attack simulation is still struggling to scale, as pointed out in earlier chapters, then it's hard to image real attacks being carried out, but as we have emphasized, the best use of these tools is often in conjunction with existing techniques and workflows. The AI doesn't have to drive the entire attack, it can be a copilot, recommending targets, warning of possible defenses, etc. While there are many ways that AI can support the offensive side, one thing is clear, its role will grow over time.

The same reward engineering used to build human-like AI agents should still be able to find new and novel attack patterns simply because of the scale of information the AI can comprehend. An adversarial training setup, with defensive and offensive agents playing against each other, can produce offensive strategies at the limits of what a human is capable of, as well as defensive strategies close to optimal. This is far afield from the inefficient game repeatedly referenced up to this point. In higher-order simulations, defenses are built purposefully against ML-driven offense, even though the offense described here is meant to simulate a human, the main difference being scale.

Malicious Actors with Malicious Intent – the AI arms race

The arms race to employ AI assistance in network analysis does not exist in a defender-occupied bubble. The future of RL and other AI technologies in the context of cybersecurity is a high-paced, large-stakes, rapidly evolving field. The race includes the potential use (and misuse) by malicious actors. As technology becomes increasingly complicated and advanced, the application in network defense will become more mature, and so will the application and expansion of cybercriminals and other threat actors using the technologies. One of the most significant potential for misuse will be automated attacks. Malicious actors could use RL techniques to automate complex attack patterns. Scanning and enumeration of a network, for example, using data collected passively through Shodan, as mentioned in Chapter 2, could be automated. Malicious actors could then feed the collected data into an AI system that continuously learns and adapts to the changing network environment, always looking for patterns indicating a new weakness or exploitable behavior.

AI has been a blessing for code developers. It can assist in code development by automatically filling out parts of the code that are repetitive, noncreative, or require standardized formatting or flow. Interactive development environment (IDE) companies have already implemented these technologies and code development tools environments like VSCode and GitHub. One of the most popular is known as Copilot (ref). This same AI-assistance technology, however, can assist malware actors in creating more sophisticated malware. Malware can adapt dynamically to its environment, changing its behavior to detect defenders' efforts and avoid detection, thus making the malware more effective at finding vulnerabilities and sniffing out the crown jewels. Even more threatening, malware can now go as far as exploiting vulnerability autonomously and pivot through network defenses without human intervention. Public code repositories release specialized AI models, trained and tuned on multiple programming languages for code creation and testing purposes, which allow any malicious actor with basic code development experience to host private "helpers" and stay off the radar. Previously complex encryption coding techniques can now be easily crafted and implemented with the help of AI. These complex coding techniques are now quickly crafted and debugged, creating encrypted exfiltration channels or file encryption (ransomware) routines. RL/ML models can be written up and tested to sift through the data returned by the malware and adapt future code to hunt and destroy even more efficiently.

Another method of data collection discussed in Chapter 2 that could be honed, refined, and used more efficiently by malicious actors using AI/ML/RL is the crafting and use of social engineering attacks. Attackers use phishing attacks to get legitimate users to leak potentially harmful information. These attacks usually occur over email, attempting to get the users to click on malicious links that install malware or backdoors or respond with non-publically available information. This information can be obviously damaging or seemingly benign tidbits like names of coworkers, network architecture, or internal company policies that can be used maliciously later when crafted and combined with other information). Thankfully, crafting a convincing-sounding email that looks the part is not easy. English structure and nuances make it a hard language for foreigners to learn. However, AI technologies have been making incredibly sophisticated advances and can now be used to craft realistic, relevant, customized emails. The technology does not stop with the written word. Attackers are now even crafting voice-based phone calls, synthetic pictures, and even faked video to make their attacks with the help of AI tools [6].

AI assistance enhances combination attacks, meaning using multiple attack vectors to achieve a goal on a target. Finely tuned social engineering attacks that place ransomware exactly where it needs to be are perfect tools for RL. Collecting and running data through RL tools to identify the most critical system or data within an organization and then encrypting it would maximize the attack's impact and the likelihood of a payout. These social engineering attacks don't stop at an organization's cybersecurity attack surface; they extend well into the realm of creating believable scams and fraud campaigns. The quickly evolving realm of AI and ML tools have allowed miscreants of all breeds to use synthetic media, from voice to video, in real time to exploit human users that are unaware of the technology's capabilities. Detection methods are being researched as quickly as possible, but with locally installed and run models and tools the battle for media legitimacy and believably will continue in the near future [7].

Zero-day development and exploitation is a lucrative business. Being on the wrong end of these attacks brings fear into the heart of any network defender. No standard defense mechanisms such as IoC blocklisting or network blocking can stop an attack with no defense or a software/hardware vulnerability that even the vendors do not know about and cannot patch or remediate. AI and ML tools can use their ability to analyze vast amounts of information and code rapidly to identify potential zero-day weaknesses. These weaknesses usually require humans to generate large amounts of testing data and cause the systems to fail in an observable manner. Machine-based help could make these vulnerabilities available to attackers before the defenders notice patterns or something is wrong. The rapid development of zero-day proofs of concepts (PoCs), combined and aided by ML-enhanced payload creation, exponentially increases the damage potential and possible avenues for exploitation. This acceleration drastically shortens the required response time for cybersecurity defenders and expands the overall attack surface. What once required specialized knowledge and years of understanding can now be achieved with enough data input and computational horsepower.

Pattern observation and pattern analysis are an ML/RL superpower. One type of attack that could utilize this strength is the employment of decoy or misdirection tactics. Suppose attackers can gather data on how defenders respond to threats by observing network changes, traffic pattern shifts, and endpoint response methods. In that case, they can feed this data into an RL model to assess the most likely outcomes. These actions by the defenders could leave other areas of the network in a less-observed state. Attackers could use AI to create sophisticated decoys or carry out misdirection tactics to confuse or mislead the defenders, making detecting, and responding to legitimate threats harder. Attackers could use the same tactics and techniques by studying how IDS and AV programs respond. Using ML to observe and model this behavior could allow attackers to craft malware or just simply hone their techniques to avoid detection. This type of modeling has been demonstrated in the past through the use of time-based evaluation and representational states of the network firewalls, while the RL algorithm is training and learning (the paper on C2 and exfiltration).

8.4 Generalization

8.4.1 Running Live

Earlier chapters have already motivated the benefits of running offline; not disrupting regular operations, being able to experiment with many options, etc. There are, however, certain things that simply cannot be done offline. Any assumptions made during offline simulations cannot be tested until an attack path is tried on a real-life network. Our recommendation for this has been to:

- Emphasize learning **across** networks, instead of training separately on each
- Further impute partial information, more intuition "baked in" to the model weights and not "hard coded" into the reward engineering.
- Take advantage of the fact that accepting partial information means simulations can run with less complete scans, outdated scans.

The methodology described thus far is trained on each network it observes individually, but this does not have to be the case. Meta-learning teaches a model how to learn, sometimes with an auxiliary model and sometimes with a more complicated architecture that optimizes a combination of individual learning and meta-learning goals. To generalize the methodology across networks, a single model must see multiple networks of various configurations, defenses, and a large amount of data. Luckily, we have already described a method for quickly building mock networks that this simulation can run on.

In RL, generalization refers to the ability of a learned policy or value function to perform well on unseen or novel scenarios that were not explicitly encountered during the training process. Generalization is a crucial aspect of RL because it enables an agent to apply its learned knowledge to a broader range of situations, improving its adaptability, and effectiveness in real-world environments. There are several ways in which generalization can be achieved in RL:

- **Function Approximation:** In many RL scenarios, the state or action spaces are continuous or very large. Instead of explicitly storing values for every possible state-action pair, function approximation techniques are used to generalize the learned values. Common methods include neural networks, decision trees, or other function approximators.
- **Transfer Learning:** Transfer learning involves leveraging knowledge gained in one task to improve learning in a related but different task. In RL, an agent may learn in one environment and then transfer its knowledge to a different but related environment. This can speed up learning in new scenarios and improve generalization.
- **Experience Replay:** Experience replay is a technique where past experiences (tuples of state, action, reward, next state) are stored in a replay buffer and sampled randomly during training. This helps the agent to break the temporal correlations in the data and promotes better generalization to diverse scenarios.
- **Curriculum Learning:** Curriculum learning involves training an agent on a sequence of tasks of increasing complexity. The idea is to first expose the agent to simpler scenarios, allowing it to gradually learn and generalize its knowledge to more complex situations.
- **Ensemble Methods:** Using ensemble methods involves training multiple models independently and combining their predictions. Ensemble methods can improve generalization by reducing overfitting to specific aspects of the training data and providing a more robust policy.
- **Policy Distillation:** Policy distillation involves training a simpler, more compact policy to imitate the behavior of a more complex policy. This can help in transferring knowledge from a complex policy to a simpler one, facilitating generalization to new scenarios.
- **Intrinsic Motivation:** Intrinsic motivation mechanisms provide additional rewards to the agent based on its own curiosity or exploration. By encouraging

the agent to explore and learn in unfamiliar situations, intrinsic motivation contributes to better generalization.

- **Sparse Rewards Handling:** Training an RL agent in scenarios with sparse rewards can lead to more effective generalization. When rewards are rare, the agent is encouraged to learn a more general policy that works across a variety of situations.
- **Hyperparameter Tuning:** Adjusting hyperparameters, such as learning rates, discount factors, and exploration strategies, can impact the generalization ability of an RL agent. Hyperparameter tuning is often performed to find a balance that promotes good generalization.
- **Domain Randomization:** In simulation-based RL, domain randomization involves introducing variability into the training environment by changing aspects like physics parameters, textures, or lighting conditions. This helps the agent generalize to a wider range of real-world scenarios.

Effective generalization is a challenging problem in RL, and the choice of algorithms, representations, and training strategies can significantly impact an agent's ability to generalize across diverse scenarios. It often requires combining the above techniques to achieve robust and adaptive RL systems.

8.4.2 Teaching Computers to Attack Computers

AI attacks persist because inherent limitations in the foundational AI algorithms provide adversaries with exploitable vulnerabilities, leading to system failures. Unlike conventional cybersecurity breaches attributed to programming or user errors, these weaknesses stem from inherent deficiencies in current state-of-the-art methods. In simpler terms, the algorithms responsible for the high efficacy of AI systems are flawed, and these systematic limitations offer opportunities for adversaries to launch attacks. Regrettably, this remains an inescapable aspect of mathematical reality, at least for the foreseeable future.

Naively, one would think that true offensive operations are not that different from a red teaming exercise, but there are some very important distinctions. A true malicious adversary usually only has one shot, plays at much higher stakes, and will use every trick in the book, no holds barred, gloves off in a real engagement. This is more than just a difference in severity, it is a difference in how risk is approached for the attacker. The largest obstacle to applying the technology described in this text to actual offense is in making it run online, which has been described in Section 8.4.1, but this is not the only modification required. To look at the path of an attacker is to analyze the attacker's strategy. Cyber warfare, like any warfare, is an arms race that is always escalating, with offense and defense always reacting to each other in a game that is never close to optimal. All tools have edge cases; over-reliance on **any** tool or **any** strategy creates a weakness; if that tool is learned, its edge cases are known and exploited. By scale alone, the attack paths and strategies of an AI attacker may be distinct from what a human would plan/execute simply because they wouldn't have the time or working memory without such a tool.

An attack enabled by AI, commonly referred to as an AI attack, involves the deliberate manipulation of an AI system with the ultimate objective of inducing malfunction. Various forms of these attacks target distinct vulnerabilities within the underlying algorithms [3]:

- **Input Attacks:** This form involves manipulating the inputs fed into the AI system to alter its output and serve the attacker's objectives. Given the fundamental nature of AI systems as complex calculative machines, influencing the input allows attackers to impact the system's output [3].
- **Poisoning Attacks:** This tactic corrupts the AI system creation process, leading to desired malfunctions. One direct method is corrupting the data used in the system's development. Modern ML relies heavily on learning from specific data sources, making data integrity crucial. Poisoning attacks compromise not only the data but also the learning process itself [3].

As AI systems become integral to critical commercial and military applications, these attacks carry severe, potentially life-threatening consequences. AI attacks serve malicious purposes through various means [3]:

- **Cause Damage:** Attackers aim to inflict harm by inducing malfunctions in the AI system. For instance, an attack may cause an autonomous vehicle to disregard stop signs, leading to collisions with other vehicles and pedestrians due to incorrect sign recognition [3].
- **Hide Something:** This involves causing a malfunction in an AI system to evade detection. An example is an attack on a content filter responsible for blocking terrorist propaganda on a social network, allowing the material to spread unchecked [3].
- **Degrade Faith in a System:** Attackers seek to undermine an operator's confidence in the AI system, prompting its shutdown. An illustration is an attack causing an automated security alarm to misclassify routine events as threats, triggering a series of false alarms that may result in the system being taken offline [3].

Despite the remarkable success of AI in the past decade, it is surprising that such attacks persist without comprehensive solutions. The focus now shifts to understanding the reasons behind the existence of these attacks and the challenges involved in preventing them.

Emergent strategies can form, where the sum is greater than the parts, and this may require new novel defenses/strategies to counter them. Inclusive of this is determining what edge cases of the tool produce non-human infeasible results. Additionally, network path considerations include determinations of invalid and out-of-date paths. Standard operating procedures that could be codified in dealing with each of these and how to best use these tools are important considerations when dealing with edge cases. Specifically, every tool and strategy has limitations and is best used when these limitations are known and understood. However, the difficulty increases as the complexity of the tool increases, partially attributed to the black-box nature of AI, which makes the sense-making of AI more challenging. The analysis of tools is powerful for both defense and offense and juxtaposes the nondeterministic and intricate nature of how AI-powered tools (in this case RL) are a "double-edged sword."

8.4.3 Where the Arms Race is Racing Toward

The perpetual escalation of cyberattacks and defense to keep pace with each other is often called an arms race. Cyber operations, both defensive and offensive, are an important, if not defining, part of modern warfare. There is, however, a broader analogy that can be made to liken cybersecurity as a whole to an exercise in naturalism, a fight for survival, of which human warfare is just a small subset. Algorithms, systems, operators, actors, nation-states, all iterating, updating, and competing in an endless battle for survival, puts cyber security in a particularly unique niche among AI applications.

Cybersecurity is the one domain of AI application that presents the conditions for true evolution. Analogies to nature in computing are as old as computer science itself. Alan Turing, in his writings [19], postulated that more advanced thinking machines would be raised like a child, undergoing a wealth of experience, similar to what we refer to as curriculum learning in modern parlance. He would go on to discuss if machines could experience taste and flavor if they could have preferences, even personalities. The conclusion of his line of thinking is the thought experiment of the Turing test, a thought experiment that is often misunderstood, leading to the idea that human-level intelligence is one that is good enough to fool us. Alternatively, there are those who believe Turing's thought experiment of the Turing test is a bit more tongue in cheek, as if to say, who cares if the intelligence is "human"? What's so great about us anyway?

Jumping 70+ years forward from Turing's visionary concepts, his ideas still play out in modern theory around AI, as do the misinterpretations of his ideas. The naturalistic analogies continue; AI as a technology is basically synonymous with the idea of artificial neural networks, named for the inspiration in the human brain. As the adoption and advancement of AI grow not just in cyber but generally in society, there is speculation about where it is all leading, and there is this idea that an important waypoint (though not destination) is human-level intelligence. Sometimes this next-level of intelligence is referred to as artificial general intelligence (AGI) or "hard AI," but it is generally defined poorly.

Machines can already outperform humans in many particular tasks: a pocket calculator from 30 years ago can still add integers better than humans can. A thumb drive can store the exact content of more books with perfect recall than a human can read (let alone recite) in their lifetime. Medical diagnostic imaging powered by computer vision can sometimes surpass human professionals. "Better than human," it seems, is a fuzzy boundary, but it's clear there are some things we still do much better, especially around eclectic, creative, holistic reasoning; exactly the kind of mental skills required of a skilled cyber operator.

While some have faith in contemporary generative AI models to eventually attain this next level of human-like intelligence, others have called out how short-sighted this prediction is. Yann LeCun, a pillar in modern AI thought and research, has pointed out repeatedly [8, 16] that the experiences of intelligence that can only read about the world that only knows a universe of tokens can't truly understand the world, and will never, for example, be able to taste strawberries as Turing teased. To have this next level of intelligence requires a worldview, a mental model of the universe that is informed by multiple senses and multiple modalities; this concept of the world and the self in it is what will lead to consciousness.

This is all a bit of fun philosophy, but what about any of this is specific to cyber domain? Hopefully, in the course of this text, we have sufficiently motivated that the challenges of cyber genuinely require this next level of intelligence, that exactly the skills we know current AI to be missing are the same critical skills of the humans that lead cyber operations today. The models presented throughout this book can be tools to help accelerate those skilled humans but to replace these humans entirely, these models need to be holistic; they need to use every piece of information available and correlate it in endless dimensions, forming a world-view. The diagram earlier in this chapter illustrating some possible interactions between several models just begins to hint at this, but as stated in the caption to that image, this is analogous to having separate thinking organs for different aspects of reality. As larger, more sophisticated models can integrate more and more of these functionalities into progressively fewer neural networks, the concepts learned by the models will be more physical and more real, and this is what empowers generalization.

The cyber domain is not unique in having difficult problems though. Economics, medicine, physical sciences, and any number of fields have hard problems, but the difficulty of the problem is just one of the ingredients that put cyber in a unique position. The cyber domain makes the environment and the world seen by the model immediately raw and accessible. An AI model for cyber security applications is an algorithm, software running on hardware, interacting with other software and hardware, sometimes interacting with other AI. In this way, Cyber is the one place where AI exists in its "natural environment." Humans studying AI focus on our analogies to human intelligence because it is the best intelligence we know (so far), but to Turing's sometimes subtle point, maybe we are not so special, maybe it is not important that advanced machine intelligence be human-like.

Even if we abandon a more strict analogy to human intelligence, Turing still seems to advise us to keep the utility of an analogy to nature more broadly. Natural forces influenced our own intelligence, through evolutionary and environmental pressures, through our own fight for survival as a species, against other competitive ecosystem challenges. Those who were better at processing and implementing this information from their environment survived to have more offspring. In the same way that our own intelligence is our minds running on brain hardware, AI models may one day be the "minds" in a different space within the cyber terrain in which they may have to compete to survive.

The world a cybersecurity model must learn is the most "world like" of any proposed application of AI aside from perhaps a robot with generalized purpose interacting with the physical world. Even then, everything a robot experiences about the physical world is through a series of sensors: audio by microphone, motion by accelerometer, vision by camera, etc. All of these signals are eventually just data in a computer somewhere, files, tables, packets, and bit streams; this is the raw form they take when they are exposed to a model. A generalized, multi-modal AI for cybersecurity, meant to understand the state, and state dynamics of computers could "build up" to understanding these more complex signals of vision or audio. In this way, the "world" a cyber AI learns is at the top of a hierarchy, under which the worlds of any other AI model are subsumed. This difference is analogous to the difference between hard coding a finite (if not very large) set of tokens to learn a language (as LLMs do today) vs. simply having the capacity to understand raw audio signals and then develop an understanding of language through experience (as every living human has ever done).

Cybersecurity defines very challenging AI problems and gives an environment that is sufficiently like a generalized "world." These are two important ingredients in defining cyber's unique role in the future of AI, but as stated, there are other hard, complex, open-ended, and multi-modal problems within cybersecurity. However, there is a third ingredient that makes cyber truly unique as fertile ground for advanced intelligence to emerge from, and that is its fundamental connection to survival. Though, AI models may have vast importance in economics, impacting the well being of millions and AI models for medicine may directly impact the mortality of many people, but these use cases are still categorically different from the impact of cyber.

To revisit Yann LeCun again, he cautions critics that we need not fear advanced AI suddenly posing an existential threat to humans because we will be designing these models ourselves and we will be the gatekeepers for how much control these models have [8, 16]. If economic or medical models cannot be controlled safely, we should not implement them, but this is not true for cyber. The nature of the arms race is that it is played out at the highest levels of international security in a competitive fashion; we only need to give control over one to potentially threaten ourselves. This by itself does not change LeCun's argument [16]; if the threat is that much bigger and only has to be given one opportunity, then we will simply give it zero opportunities.

Unfortunately, this is not how it may play out. In popular science fiction, depictions of an AI take over, the intelligence gains command of nuclear capabilities, often leading to a doomsday scenario, either because the AI gained control unintentionally or because it was explicitly handed control because the AI was trusted. In reality, a much more likely scenario is that AI will be given control of such systems not because of trust or mistake but because of necessity, because of the arms race. If AI is capable of coordinating attacks independently and having defensive responses also independently (as this book has laid the earliest foundations for), then the ultimate end state becomes clear: the question is not whether we put our lives in human hands or the hands of AI, but rather, do we put our lives in the hands of our AI or the AI of our enemies. If we do not trust our own AI to coordinate a defensive strategy, then our most critical systems will be in the control of an enemy AI that can beat our most capable humans. To be a complete "Luddite" in our defense, to completely unplug and isolate our own capabilities is unrealistic: all systems are vulnerable, no defense is perfect, and to downgrade our defenses to the extent that AI is irrelevant as an attack vector would just hobble ourselves in another way.

This all sounds very grim; why even pursue AI if we think this is the inevitable conclusion? The answer has two parts: first, it does not matter if we pursue it, someone will. This is inevitable. In analogy to the nuclear arms race, regardless of espionage efforts to accelerate world powers gaining nuclear capabilities, independent groups of scientists were coming to the same advances at nearly the same time, and advanced AI will be the same. Like the nuclear arms race, a large incentive to participate in the AI arms race is that the mere fact that I own the same technology is a deterrent for you to attempt to use yours on me. The second part, more optimistic than the first compulsory part, is that we should pursue AI because it is still the brightest of possible human futures, even with the risks.

If we cannot figure out advanced AI, our technological advances in general will stagnate. We as a group, are not smart enough, we can only scale so far, we have too many problems. We cannot even agree on how to gather and distribute food among ourselves. Our brightest are too few, and still not bright enough. The way out, the way forward, is AI. It's a realm of endless possibilities, a frontier waiting to be explored. Just as nuclear physics led to advancements in medicine and energy, AI has the potential to revolutionize every aspect of our lives, offering new solutions to our most pressing challenges.

AI will be like this, but touching every aspect of our lives. Based on the arguments listed above, it seems probable that cybersecurity will be the first place humans are forced to turn the keys over to an AI that truly surpasses us. So, it is upon us to design such a system with care so that it will serve us well, even if it is not human-like.

Despite some nuanced disagreement with LeCun's outlook for AI, the authors of this work share his optimism, and share his view on what our course of action should be, to build these intelligent systems responsibly. Turing describes training a machine intelligence like raising a child, and that is how we should think of it, as our descendent, our inheritor, who we put our hopes and dreams upon, but who we know will have their own hopes and dreams as well. Similar to how our children may be more progressive in their outlook compared to their conservative parents, it is our duty to ensure that the AI we create is not just intelligent, but also ethical and responsible, reflecting our values and aspirations.

References

1 https://github.com/BloodHoundAD/BloodHound?tab=readme-ov-file#readme.

2 Fadwa Alrowais, Sami Althahabi, Saud S Alotaibi, Abdullah Mohamed, Manar A Hamza, and Radwa Marzouk. Automated machine learning enabled cyber security threat detection in internet of things environment. *Computer Systems Science and Engineering*, 45(1):687–700, 2023.

3 Marcus Comiter. Attacking artificial intelligence: Ai's security vulnerability and what policymakers can do about it. Technical report, Harvard Kennedy School Belfer Center for Science and International Affairs, August 2019.

4 Arun Kumar Dey, Govind P Gupta, and Satya P Sahu. A metaheuristic-based ensemble feature selection framework for cyber threat detection in IoT-enabled networks. *Decision Analytics Journal*, 7:100206, 2023.

5 Akshat Gaurav, Brij B Gupta, and Prabin K Panigrahi. A comprehensive survey on machine learning approaches for malware detection in IoT-based enterprise information system. *Enterprise Information Systems*, 17(3):2023764, 2023.

6 Craig Gibson and Josiah Hagen. Virtual kidnapping, June 2023. URL https://www.trendmicro.com/vinfo/es/security/news/cybercrime-and-digital-threats/how-cybercriminals-can-perform-virtual-kidnapping-scams-using-ai-voice-cloning-tools-and-chatgpt.

7 Will Jackson. From face-swapping to chatbots - here's how South-East Asia's scammers are using AI tools. *ABC News*, May 2024. URL https://www.abc.net.au/news/2024-05-16/pig-butchering-scams-artificial-intelligence-ai-face-swapping-/103804830.

8 Stephen Levy. How not to be stupid about AI, with Yann Lecun. *Wired*, 2023. URL https://www.wired.com/story/artificial-intelligence-meta-yann-lecun-interview/.

9 Xiang Ling, Lingfei Wu, Jiangyu Zhang, Zhenqing Qu, Wei Deng, Xiang Chen, Yaguan Qian, Chunming Wu, Shouling Ji, Tianyue Luo, et al. Adversarial attacks against windows PE malware detection: a survey of the state-of-the-art. *Computers & Security*, 128: 103134, 2023.

10 Infosecurity Magazine. Infosecurity podcast episode 44, 2023. URL https://www.infosecurity-magazine.com/podcasts/podcast-episode-44/.

11 Routa Moussaileb, Nora Cuppens, Jean-Louis Lanet, and Hélène Le Bouder. A survey on windows-based ransomware taxonomy and detection mechanisms. *ACM Computing Surveys (CSUR)*, 54(6):1–36, 2021.

12 Sathvik Murli, Dhruv Nandakumar, Prabhat Kushwaha, Cheng Wang, Christopher Redino, Abdul Rahman, Shalini Israni, Tarun Singh, and Edward Bowen. Cross-temporal detection of novel ransomware campaigns: a multi-modal alert approach. *arXiv:2309.00700 [cs.CR]*, 2023.

13 Dhruv Nandakumar, Robert Schiller, Christopher Redino, Kevin Choi, Abdul Rahman, Edward Bowen, Marc Vucovich, Joe Nehila, Matthew Weeks, and Aaron Shaha. Zero day threat detection using metric learning autoencoders. In *2022 21st IEEE International Conference on Machine Learning and Applications (ICMLA)*, pages 1318–1325, 2022. doi: 10.1109/ICMLA55696.2022.00210.

14 Dhruv Nandakumar, Devin Quinn, Elijah Soba, Eunyoung Kim, Christopher Redino, Chris Chan, Kevin Choi, Abdul Rahman, and Edward Bowen. Foundational models for malware embeddings using spatio-temporal parallel convolutional networks. *ArXiv*, abs/2305.15488, 2023. URL https://api.semanticscholar.org/CorpusID:258888134.

15 Harun Oz, Ahmet Aris, Albert Levi, and A Selcuk Uluagac. A survey on ransomware: evolution, taxonomy, and defense solutions. *ACM Computing Surveys (CSUR)*, 54(11s):1–37, 2022.

16 Billy Perrigo. Meta's AI chief Yann Lecun on AGI, open-source, and AI risk, 2024. URL https://time.com/6694432/yann-lecun-meta-ai-interview/.

17 Salwa Razaulla, Claude Fachkha, Christine Markarian, Amjad Gawanmeh, Wathiq Mansoor, Benjamin C M Fung, and Chadi Assi. The age of ransomware: a survey on the evolution, taxonomy, and research directions. *IEEE Access*, 2023.

18 Mannepalli Sravanthi, Gundeti Suchithra, and Pavuloori Vennela. Cyber threat detection based on artificial neural networks using event profiles.

19 Alan Turing. Computing machinery and intelligence. *Mind*, LIX:433–460, 1950. URL https://academic.oup.com/mind/article/LIX/236/433/986238.

20 Cheng Wang, Akshay Kakkar, Chris Redino, Abdul Rahman, S Ajinsyam, Ryan Clark, Daniel Radke, Tyler Cody, Lanxiao Huang, and Edward Bowen. Discovering command and control (C2) channels using reinforcement learning (RL). *Submitted IEEE SouthEastCon 2023*, 2023.

21 Sergio Caltagirone, Andrew Pendergast, and Christopher Betz. The diamond model of intrusion analysis. *Threat Connect*, 298(0704): 1–61, 2013.

22 MITRE. Mitre ATT&CK framework, 2023. URL https://attack.mitre.org.

9

Model-driven Penetration Testing in Practice

9.1 Recap

As we have discussed in previous chapters, this book's goal was to discuss how to leverage the attack graph construct [23] and train RL models over them to predict weaknesses within network topologies to identify the most likely locations (or paths) where adversaries could perform both pre- and post-exploitation activities (e.g., exfiltration). Previous work, however, constructs attack graphs either with no vulnerability information [3, 13, 14, 35] or entirely with vulnerability information [5, 17, 41].

This perception of the inadequacy of the attack graph concept hampers the progress of automated cybersecurity testing. Recent advancements in reinforcement learning (RL) seek to redefine the generation of attack graphs as a cybersecurity best practice. As discussed in earlier chapters, RL focuses on interactive learning by training agents to develop optimal policies that link states to actions, optimizing Markov decision processes (MDPs) within the network models (or attack graphs themselves). Unlike exhaustive attack graphs, these optimal policies enable the identification of the most critical individual attack paths.

Contemporary automated penetration testing employs rule-based methodologies and model-checking principles to systematically explore potential network attacks, pinpointing instances that breach correctness or security protocols by constructing an attack graph. While these top-down approaches strive to encompass all conceivable attacks, they inadvertently struggle to discern the most crucial ones. Instead of identifying the proverbial needle in the haystack, these methods often proliferate more haystacks or haystacks containing needles. The challenge intensifies in emerging network landscapes, such as 5G and the Internet of Things (IoT), where networks may consist of thousands of hosts, continually evolving over time.

Reinforcement Learning for Cyber Operations: Applications of Artificial Intelligence for Penetration Testing, First Edition. Abdul Rahman, Christopher Redino, Dhruv Nandakumar, Tyler Cody, Sachin Shetty, and Dan Radke.

This chapter delves into the technical shortcomings of existing methods that propel the adoption of RL. A critical examination of RL and MDP-based approaches for automating penetration testing is discussed, followed by an overview of recent research in RL for penetration testing. This chapter identifies ongoing research trends and advocates for "whole campaign emulation" as a long-term objective for RL-based penetration testing automation. Future directions for research are also discussed.

Chapter 3's discussion presented RL as an agent that interacts with an environment across discrete time steps, selecting actions at each step. The environment responds by providing the agent with a new state and a reward. This interaction forms a sequence of states and actions until the agent reaches a terminal state (e.g., gaining escalated privileges on a target host). Formally, the states, actions, admissible state-action pairs, state transition probabilities, and rewards constitute the MDP. RL agents through their interaction with MDP environments, *learn* to minimize the discounted sum of expected future rewards (see Figure 4.1).

In earlier chapters, common vulnerability scoring system (CVSS) scores were used to provide an empirical and automatic means of constructing attack graphs for RL but do not always correlate to a useful contextual picture for cyber operators. By relying totally on abstractions of CVSS scores, network representations, unfortunately, can be biased totally toward vulnerabilities only and not on a realistic view of how an adversary plans or executes their attack campaign. Chapter 2 discussed the role of CVSS scores to construct attack graphs [5, 9, 17, 41] that converge to unrealistic attack campaigns. However, the approaches described in this book have led to substantial improvements as outlined as follows. We integrated OAKOC cyber terrain [7] into MDP models for attack graphs [11]. Additionally, we applied this methodology to formalize an AI driven method for penetration testing/red teaming using reinforce learning (RL), designating specific devices (e.g., firewalls) as elements of cyber terrain OAKCOC (e.g., obstacles) within a representative network that is at least ten times larger than those employed by previous researchers [3, 5, 13, 14, 17, 35, 41]. We contributed to the literature by utilizing CVSS scores to construct attack graphs and MDPs, introducing RL in crown jewel analysis (CJA-RL). This approach incorporated network structure, cyber terrain, and path structure to streamline operator workflow, automating the construction, and analysis of attack campaigns from various entry points or initial nodes to a CJ's 2-hop network [11].

We introduced a methodology for modeling service-based defensive cyber terrain in dynamic attack graph models. For instance, in [6], we developed an RL-based algorithm for identifying the top-N exfiltration paths in an attack graph was proposed [6]. In our surveillance detection route using RL (SDR-RL), we proposed a method that employs a warm-up phase and penalty scaling to balance the asymmetry between the number of host services scanned, and the volume of defensive terrain encountered [18]. Our work mirrors the asymmetry sought by human operators during network reconnaissance in general and SDR in particular. We extended the double agent architecture (DAA) introduced by Nguyen et al. [30], originally based on standard deep Q-learning (DQN), by incorporating actor-critic (A2C), and proximal policy optimization (PPO) algorithms [18, 34]. This was a critical contribution as it lays the groundwork for multi-objective approaches [19] along with future methods for more robust AI-enhanced penetration testing methods.

The following sections review how the RL developed in this book impacts the detection of adverse behavior for various cybersecurity areas that include cyber terrain, crown jewels discovery, analysis of exfiltration paths, and advanced reconnaissance pathway discovery (i.e., surveillance detection routes).

9.1.1 Using Cyber Terrain

While CVSS scores provide a strong foundation for attack graphs, we posit that notions of *cyber terrain* [7] should be built into attack graph representations to enable RL agents to construct more realistic attack campaigns during penetration testing. In particular, we suggest a focus on the OAKOC terrain analysis framework that consists of obstacles, avenues of approach, key terrain, observation and fields of fire, and cover and concealment [7].

9.1.2 Crown Jewels

Cyberattacks pose existential, nation-level threats to electrical and financial infrastructure [15, 37], thereby directly challenging societal stability and domestic prosperity. Importantly, cyberattacks, such as the infamous Colonial Pipeline Hack [32], target domestic prosperity by way of enterprises [37].

Enterprises and organizations generally depend on key information technology (IT) systems, known as crown jewels (CJs), operating as designed. The criticality and importance of CJs can be described in terms of the function and data they host [28]. According to MITRE [26], "Crown Jewels Analysis (CJA) is also an informal name for mission-based critical information technology asset identification. It is a subset of broader analyses that identify all types of mission-critical assets." Commonly, adversaries compromise CJs not by way of ingenious technical methods, but rather by the apt use of cyber terrain [7]. Adversaries often gain access to key terrain slowly over time [16].

There has been a growing interest in using notions of cyber terrain in cyber engineering practice [2, 7, 16]. Chief among the reasons is an increasing appreciation for the integral roles that networks, their structure and configuration, and the paths taken in them play in engineering for cybersecurity, deterrence, and cyber resilience. While the MITRE ATT&CK framework [36], for example, provides a temporally organized compendium of tactics and techniques used by adversaries, it is not clear that attack campaigns and defensive measures can be well engineered by selecting and aggregating elements from the framework. An ordered list of techniques may provide a meaningful post-hoc accounting of attack campaigns. But, for most operators, information regarding a network's structure, e.g., as captured by an attack graph, an understanding of a network's cyber terrain, and the overall path structure within the network are foundational to operational application of MITRE ATT&CK and similar frameworks.

Though cyber terrain has been advocated for [2, 7, 16, 20], it has not been widely applied to automated methods like machine learning. Gangupantulu et al. present methods for building cyber terrain into MDP models of attack graphs [11]. Specifically, firewalls are treated as cyber obstacles[1].

Typically, mission assurance and resiliency processes start with CJA as a methodology to align an organization's mission with their critical cyber assets [26]. Prior to initiating such an analysis, the critical needs of the business mission must be defined. These findings can be employed to scope and size the infrastructure required to fulfill the mission need. The IT assets that align to this mission are considered to be the most important and of the highest value; they are the "Crown Jewels" of the organization.

The assignment of a mission relevance value for these IT assets lies at the heart of CJA [26]. The depth of such an analysis depends on the functional use of the IT asset by the organization. For example, a payroll system may be a target as it has not only information related to salaries but highly sought after personal identifying information (PII) in a single location. Such data, once exfiltrated, can be sold at a high price or used for nefarious purposes. By measuring the importance of these IT assets, the organization is capable of assigning risk and value scores for each, and CJs can be determined computationally [16].

In real-world cases of penetration testing and adversary emulation, attackers do not always follow direct or expected attack paths [4, 40]. The assumption that attackers will follow direct paths when performing CJA creates gaps in assigned risks and values. While MITRE's ATT&CK Framework highlights methodologies used to exploit these gaps [36], organizations struggle to derive the secondary or tertiary effects of these exploits. As a result, such gaps create a situation where reconnaissance and access exploits during late-stage attacks can appear to have little impact on an organization's CJs, when in reality adversaries are using them to position themselves near key terrain and avenues of approach to further their attack campaign goals. Consider the following example of such a phenomena.

1 Obstacles are part of the OAKOC intelligence preparation of the battlefield framework described in [7].

9.1.3 Exfiltration

The National Institute of Standards and Technology (NIST) special publication 800-53 revision 5 states that exfiltration[2] (also called exfil) is the unauthorized movement of data within a network [29]. Many times, cyberattacks are considered successful if they exfiltrate data for monetary, disruptive, or competitive gain. Detection of exfiltration can be plagued with technical challenges as adversaries routinely encapsulate data within typically allowable protocols (e.g., http(s), DNS) which make it significantly harder to defend. Additionally, adversaries have been known to prefer traversing certain network paths for data theft to reduce detection and tripping cyber defenses so they do not raise suspicions.

Heisting data requires two different plans: a plan to get to the data and a plan to exfiltrate the data without getting caught. Much effort in the cybersecurity industry is devoted to identifying and preventing points of weakness that allow authorized (i.e., adversarial) entry into a network. The most common exfiltration opportunity is moving data from a local network to an adversary network via the internet. To perform this, an adversary must gain access to the data on an organization's network, then send the data to a place off their network. Most organizations are focused on preventing network access, which leaves gaps in defenses for access from the network to the internet.

Much of the literature on automating penetration testing using RL has a focus on the way networks can be accessed (i.e., infiltration [3, 5, 10, 11, 13, 14, 17, 35, 41]). And while some consider using RL to detect exfiltration [1, 38], RL for conducting post-exploitation activities like exfiltration are under-studied [21]. Maeda and Mimura [21] apply deep RL to do exfiltration; however, they do not use a standard attack graph construct, but rather define states using an ontological model of the agent and define actions using task automation tools [21]. Their approach has several limitations:

- The RL agent's inputs and outputs are greatly abstracted away from network structure, path structure, and cyber terrain, thereby limiting the ability to anchor agents to the *real* computer network.
- The exfiltration methodology does not leverage automated frameworks for attack graph construction like MulVAl [31] or the vulnerability- and bug-reporting communities (e.g., via the CVSS [24]).
- The output of the RL-based exfiltration method is not easily interpretable in terms of networks, their paths and configurations, and risk preferences regarding their traversal.

2 NIST 800-53r5 [29] states specifically that exfiltration lies within security control SC-07(10) for boundary protection to prevent unauthorized data movement (exfiltration).

9.1.4 Surveillance Detection Routes (SDR): Advanced Reconnaissance

Reconnaissance (also called recon) in MITRE's Adversarial Tactics, Techniques, and Common Knowledge (ATT&CK) framework is described as "techniques that involve adversaries actively or passively gathering information that can be used to support targeting [25]." As reconnaissance activities usually precede an exploitation campaign, detection of these efforts benefit cyber defenders by identifying potential targets (e.g., crown jewels) of interest. In this respect, adversarial recon activities strive to maximize visibility of targets while minimizing opportunities of being detected. Critical to this are the identification of paths, termed SDR, traversed by adversaries to gather critical data about targets (e.g., ports, protocols, applications, services).

From a cyber protection standpoint, recon activities disguise serious hostile intent but may be quite difficult to quantitatively differentiate from normal behavior. Malicious intent is quite difficult to observe, as it may be efficiently designed to hide in the background of normal traffic. Detecting this type of recon is situational in nature and requires meticulous analyses of huge volumes of collected data. Other domains apart from cyber are often challenged in a similar manner, where evaluation of these SDR require data analysis to differentiate abnormal traffic from events that are suspicious in proximity to roads and crossings [33].

Modern efforts to detect and respond to adversarial network reconnaissance are a complicated blend of automated and human processes. Automated collection systems are installed on network devices and endpoints to passively and actively monitor the network communications, analyze the flow, and aggregate the data for the security information event management (SIEM) and/or security orchestration, automation and response (SOAR) systems for analysis. These network tools assist the human component of detection by providing automated security reports, incident alerts, and executing network protection protocols with a single click from the security operations center (SOC) analyst. The effectiveness of these systems relies on the data collected, the knowledge of current threat behavior, and the human analyst's ability to understand the threat. Naturally, such approaches have blindspots.

Combining the current security information (network topology and configuration) with machine learning (ML) analysis allows highlighting weak points missed by automated systems that an attacker may focus on during initial recon. Network traffic behavior analysis, no matter how advanced, relies on active network traffic and does not preempt network/host/protocol misconfigurations. This paper contributes a deep RL approach to generating SDR in the form of attack graphs from network models consisting of network topology and configuration, thereby extending the suite of automated tools and systems available for the cyber defense.

9.2 The Case for Model-driven Cyber Detections

Reinforcement learning (RL) in penetration testing holds promise in addressing numerous challenges, as highlighted by the inadequacy of a single penetration testing tool. RL serves as the foundation for diverse tools, encompassing analysis, security bypassing, and penetration across various testing types, including external, internal, blind, and double-blind testing. The automation and generality of RL enable its swift deployment in various forms, each with distinct policies, at multiple points within a network. This versatility becomes increasingly crucial in the era of the Internet of Things, where scalable intelligent payload mutation and entry-point crawling, tasks well-suited for RL, become essential in penetration testing.

The accelerated pace of simulations compared to the evolution of real networks allows RL agents to iteratively optimize toward network specifics. In simulation, when a change occurs, the simulator is promptly updated, enabling the RL agent to adapt to new conditions efficiently. Consequently, RL agents learning in simulation tend to exhibit a more specialized intelligence, focusing on specific network scenarios without the need for broad generalization. In contrast, in real networks, RL agents must cope with the network's evolution, necessitating a higher degree of general intelligence.

RL for penetration testing commonly adopts the attack graph model, treating the environment as either a MDP or a partially observable Markov decision process (POMDP). While POMDPs offer more realism, their scalability to large networks remains unproven, necessitating the modeling of numerous prior probability distributions. MDPs, despite their simplification, allow for scalability and a more practical worst-case analysis, making them the risk-averse option prone to false alarms. The chosen MDP framework is employed to facilitate scalability and future extensions to POMDPs through the incorporation of cyber terrain.

Unlike most prior RL work in penetration testing, this approach incorporates vulnerability information, utilizing the CVSS. Additionally, it extends beyond vulnerability information by integrating concepts of cyber terrain. Drawing from the literature, the deep q-network (DQN) is utilized as the RL solution method, with the distinction of employing a larger network than those previously reported.

There is a distinction between learning in simulation and learning in real networks concerning the timing of disruptions and their effects. Learning in simulation entails mistakes with a lower impact compared to learning in real networks, where errors can have more severe repercussions. Simulation-based learning is also less disruptive to the tested network. At a fundamental level, the rate at which the RL agent interacts with and learns from the network model is constrained by computational resources in simulation, whereas in real networks, it is bounded by actual network processes.

Learning in simulation proves practical, minimally intrusive, and appears achievable with current state-of-the-art RL algorithms. This dichotomy between learning in simulation and real networks is characterized by differences in consequence severity, network disruption, rate of interaction/learning, and the intelligence required – narrow intelligence for simulation-based learning and general intelligence for real networks.

It is noteworthy that methods for learning from real networks offer a direct connection of RL agents to the actual system, reducing biases introduced by using abstractions like network models, attack graphs, and related MDPs. This direct connection theoretically allows for a less engineered reward signal and, consequently, fewer designer-imposed performance limits. This stands in contrast to learning attack graphs using simulations, where significant attention is dedicated to reward engineering.

9.2.1 The Environment

The cornerstone of RL is a well designed environment specifically suited for achieving the target objecting. In our cyber context, this environment would be a detailed digital representation of the target host network, replete with host information such as operating systems and available services, networking information such as firewalls, and host vulnerability information. The goal of creating such a rich virtual environment is to allow the AI agents to replace human operator decisions as closely as possible by providing them as much of the same information as possible.

9.2.2 The CVSS Attack Graph

We construct our environment by first creating an attack graph of the target network. That is, a digital graph consisting of all the networked assets on the network, the nodes, and all the possible connectivity configurations between those hosts, the edges. Consequently, this graph represents all of the discoverable assets on the network which the AI agents can surveil or use as intermediate destinations on their way to a target host. This host representation can be created using a NMAP scan of the target network.

The next step is to layer vulnerability information of the hosts on the network onto the attack graph. The purpose of adding vulnerability information is to be able to engineer rewards for the AI agents as they traverse the environment. Like their human counterparts, AI agents will be trained to prefer traversing more vulnerable hosts by providing them with higher transition probabilities to those hosts, as well as higher rewards for moving to them. These probabilities and rewards are computed by computing the CVSS score of each asset. Particularly, the transition probabilities P between hosts can be assigned using the CVSS Attack Complexity score. Intuitively, a more complex exploit will have a lower likelihood of succeeding, and consequently a lower transition probability. For our applications, transition probabilities assume three distinct values: 0.9 for high probability, 0.6 for medium probability, and 0.3 for low probability attacks. Furthermore, rewards for exploiting hosts can be assigned using the CVSS Exploitability Score and can be computed using the formula:

$$Reward = Base\ Reward + \frac{Exploitability\ Score}{10} \tag{9.1}$$

where the agent is given a base reward of −1 for every step it takes, and +100 for reaching the target host. The negative reward allows the agent to minimize it's footprint while encouraging to reach the target node using the large positive reward.

9.2.3 Layering Defensive Terrain

Now that we have constructed an attack graph containing contextual information about vulnerabilities in the environment, our next step is to add an additional layer of context apropos of the already existing defensive terrain present in the environment. The defensive terrain could include firewalls, DMZs, or security software such as endpoint monitoring tools. Layering such information into the simulated environment will allow the AI agent to approximate real-world considerations of risk of detection during exploits without sole focus on reward for exploitation. Consequently, given that such defensive terrain is meant to serve as a "counterweight" for vulnerabilities, they are modeled as negative rewards for every action taken by the agent. In keeping with the theme of closely approximating human penetration testing scenarios, we do not explicitly define defensive terrain into the states of the attack graph. Instead, we allow the agents to infer the presence of such defenses based on the services available on each individual host such as host-based antivirus. We engineer defensive terrain to be additive, or iterative layered on top of vulnerability based rewards. The negative reward for defensive terrain can be computed based on two key factors:

1) A risk hierarchy applied to service categories. That is, some categories of services are inherently riskier to exploit than others.
2) The type of exploitative action performed by the agent on a service.

Negative reward can be computed for each unique service-action pair and added to overall reward during training. Our reward hierarchy assigns ?6 for authentication, ?4 for data, and ?2 for security and common services. Moreover, the type of actions has an effect on the rewards. We add +1 reward for (?1, ?3, and ?5) scanning actions while n (?2, ?4, and ?6) is assigned for exploiting actions.

9.2.4 The AI Agents

Now that the simulated cyber environment has been constructed, we can now begin to think about the agents that we will implement to learn it. One key design consideration while doing so is the size of the environment. The size of the state and action spaces in the environment is greatly determined by the number and diversity of the hosts in the simulated environments. Therefore, we must select or create agents and algorithms that can appropriately adapt to enterprise scale networks. Nguyen et al. [30] proposed a novel framework that splits the task of learning this state-action space to two distinct neural networks: the structuring agent and the exploiting agent. Splitting the space to two agents allows for easier, more stable learning of the network. The following sections will explore both these agents as well as the algorithm used to train them.

9.2.5 The Structuring Agent

The structuring agent is responsible for learning structural information about the environment. Such information can include information about hosts present in a network, subnets, and firewalls. Essentially, the structuring agent is tasked with discovering any physical or logical assets on a network – the CVSS attack graph.

The structuring agent is a feed-forward neural network that begins its learning process by observing the environment and retrieving a state vector from the environment as input. The agent then uses this information to produce an action, which in this case can belong to one of two categories:

- **A scanning action:** The agent thinks that the observed state lacks structural information, or that it can explore it for more deeply.
- **An exploiting action:** The agent thinks it has all the structural information it needs for an exploit to be performed.

In the case of a scanning action, the agent returns a scanning action to the environment, such as host or subnet scanning, and is returned an immediate reward for the action which it can learn. In the case of an exploiting action, it will trigger the exploiting agent to make a decision on what exploit should be performed.

9.2.6 The Exploiting Agent

The exploiting agent is triggered if the structuring agent feels that the current state is best suited for an exploitative action. The structuring agent will provide the exploiting agent with a state vector containing the selected host, reachability information, and the services running on a host. The exploiting agent will then make a decision for what kind of exploit it wants to execute, and this can include scanning actions. The selected exploit will then trigger a reward from the environment which the exploiting agent can use to update its policy. This reward is also sent to the structuring agent which will update its policy accordingly.

If the structuring agent does not trigger the exploiting agent, the reward from the exploiting agent is considered 0, and the structuring agent updates is policy solely based on the reward earned from its performed scanning action.

9.2.7 The Intuition

Delegating the learning of the environment to two distinct agents allows the environment to be more tractable, by reducing the space each agent has to learn significantly. If only one agent is used to learn the environment, the state-action space enumeration would be exponentially larger and consequently, it would take a much longer time to train the agent if at all. Using a multi-agent strategy allows us to save valuable training time while simultaneously allowing our agents to learn much more. Furthermore, the independence between agents allows for independent training if necessary. This comes in handy particularly in scenarios where hosts are added to networks, or network structure changes but the services running on hosts do not, in which case the structuring agent can be trained without retraining the exploiting agent.

9.2.8 The Training Algorithm

The structuring and exploiting agents are trained using a PPO algorithm which generally provides enhanced stability and reduced training as compared to traditional RL approach such as deep-q learning or advantage actor-critic methods. Before discussing the algorithm, it's first important to understand why it's needed.

Conventionally, DQN is the most basic algorithm used for modeling RL on penetration testing. Nguyen at al. Nguyen et al. [30] proposed a method that utilizes two A2C agents: one called the structural agent that learns the structural information about subnets, hosts, and firewalls as well as the connections between them. The other called the exploiting agent selects actions to take and the target of them. Their method solves the scalability problem of DQN to some extent and has a better capacity for large networks. Recall that one of the pitfalls of traditional RL methods is that once your agent/actor (depending on the algorithm) starts to learn a bad policy, it is very difficult for it to recover during the training process. One way to address this is by reducing the impact of policy updates between training episodes, which essentially stabilizes training between episodes. Think of this as akin to setting a learning rate for policy updates. PPO does exactly that by computing the ratio between the newly updated policy and the older policy. We can then use this ratio to determine how much change in the older policy we are willing to tolerate. This percent change can be used to update the actor networks of each agent separately. As mentioned earlier, we improve upon this method by applying PPO instead of A2C for both of the agents. PPO is an advanced RL algorithm known for its convergence speed, stability, and sample efficiency. It optimizes the following clipped surrogate objective function to prevent performance collapse caused by a large policy update:

$$\mathcal{L}(\theta) = \mathbb{E}\left[\min\left(\rho_t(\theta)A_t, \text{clip}\left(\rho_t(\theta), 1-\epsilon, 1+\epsilon\right)A_t\right)\right], \tag{9.2}$$

where $\rho_t(\theta) = \pi_\theta(a_t|s_t)/\pi_{\theta_{\text{old}}}(a_t|s_t)$ is the probability ratio of the new policy over the old policy. The advantage function A_t is often estimated using the generalized advantage estimation [34], truncated after T steps:

$$A_t \approx \delta_t + (\gamma\lambda)\delta_{t+1} + \cdots + (\gamma\lambda)^{T-t+1}\delta_{T-1}, \tag{9.3}$$

$$\text{where } \delta_t = r_t + \gamma V(s_{t+1}) - V(s_t). \tag{9.4}$$

9.2.9 Learning in Simulation vs. Learning in Reality

A brief distinction is made between learning in simulation and learning in real networks. There is an immediate difference with respect to disruptions and consequences. Compared to learning in real networks, mistakes made when learning in simulation have a lower consequence. Also, learning from simulations is less disruptive to the tested network. At a deeper level, in simulation, the rate at which the RL agent can interact with and learn from the network model is bounded by access to compute, whereas the rate at which the RL agent can interact with and learn from the real network is bounded by real network processes.

Since simulations can be run much faster than the network evolves, when learning in simulation, RL agents can iteratively optimize toward network specifics. That is, when a change occurs, the simulator is updated, and the RL agent is reoptimized for the new conditions. As a result, RL agents that learn in simulation can have comparatively more *narrow* intelligence in the sense that they do not have to be able to perform well in networks generally. In stark contrast, in real networks, the RL agent cannot learn dramatically faster than the network evolves and therefore the agent must generalize across network specifics. Thus, learning in real networks requires significantly more *general* intelligence.

For these reasons, learning in simulation is practical, minimally intrusive, and seemingly achieveable with existing state-of-the-art RL algorithms. Moreover, for these reasons, one can distinguish learning in simulation and learning in real networks as two different activities, characterized by remarkably different learning settings: low vs. high consequence mistakes, low vs. high network disruption, low vs. high rate of interaction/learning, and narrow intelligence vs. general intelligence, respectively.

Note, methods for learning from real networks allow for a direct grounding of RL agents to the real system, i.e., it reduces biases and distortions created by using abstractions like network models, attack graphs, and related MDPs. In theory, this should allow for a less engineered reward signal, and, in turn, less designer-imposed limits on performance. In fact, much of the literature on learning attack graphs using simulations is concerned with reward engineering.

9.2.10 Putting it in Practice

The following sections will explore an implementation of the above approach to identify surveillance detection routes (SDR) on a small, but realistic, simulation of an enterprise network. The goal is to demonstrate how the previous chapters can be integrated to provide cyber analysts with relevant intelligence.

9.2.10.1 The Motivation and Experimental Design

The motivation of the following modeling approach is to identify possible hosts and routes malicious actors can utilize to gain maximum surveillance exposure at different points within the simulated network. That is, gain as much insight into a network's topology, connectivity, defensive terrain, and host information while minimizing surveillance detection. As discussed in previous chapters, RL techniques require agents to have a tangible "end state," i.e., a natural stopping point for when the agent can stop the training episode and start again – this is usually the point where the agent gains the maximum reward. For the purposes of discovering SDRs, we will identify a single host on the simulated network as the terminal node. Discovering the terminal node will mark the natural end of a training episode and the agent will receive a +100 reward for doing so. We will also choose different starting hosts/nodes at random during training runs. The objective of agents during training will be to maximize the gained reward by maximizing network discovery while reducing its exploitative exposure and risk of discovery.

9.2.10.2 Network Design, Assumptions, and Defensive Terrain

The experimental network where the simulations are run was derived from an architectural leading practices approach to represent enterprise network configurations and deployments. The network contains:

- Defined Subnets – 9
- Defined Hosts – 26
- Types of Operating Systems – 2
- Privilege Access Levels – 3
- Network Services – 9
- Host Processes – 6
- Network Firewall Rulesets – 39

Subnets are constructed to represent a grouping of hosts with commonly segregated services utilized for enterprise information technology administration to include server services, database services, client workstation networks, edge and DMZ services, and core services that orchestrate least privilege or zero-trust security (i.e., domain controllers and public-key infrastructure). The services running on the hosts in the individual subnets are assigned based on the following presumptions which as derived from enterprise security best-practices [18]:

- Authentication services are exposed to the internet through a virtual private network (VPN).
- Web services are exposed to the internet through a secured edge network zone (DMZ).
- Services exposed to the internet are monitored.
- Firewalls are monitored at a higher rate than other network devices.
- Security services have the most inherited security controls.
- Authentication services and firewall services, if exploited, have the greatest secondary and tertiary impacts on a network's overall security profile.
- Network security rules only apply allowlists.
- Host and network assets apply principles of least privilege when authorizing privileges for account access and use.

The services available on hosts can belong to one of four categories: authentication, data, security, or common. Performing a scanning action on a service returns a negative reward of ?5 for authentication, ?3 for data, and ?1 for security and common. Performing an equivalent exploit will return rewards of ?6, ?4, and ?2 respectively. This is because we consider exploitative actions to be inherently riskier than scanning actions. Furthermore, in order to account for individual organizational risk tolerance, we can also scale the negative reward using a penalty factor, which is basically a constant value by which we multiply the negative reward. The higher the penalty factor, the more negative reward for risk actions, and consequently, the more risk-averse the agents will be.

At this point, you might be wondering why an agent would ever need to perform an exploit if network discovery can, theoretically, be accomplished by less risky scanning actions. This is because we have engineered our simulated network as realistically as possible such that certain actions can only be successful based on the services running on a host and the assumed identities of agents. For example, in certain parts of our network, scanning actions can only be performed after authentication, etc. In terms of defensive terrain, we configured our network as follows.

To simulate real-world network conditions, there are layers of defenses between the Internet and the innermost private network. Systems that require Internet to provide service (HTTP, email) are most vulnerable to attack; and are typically in the DMZ (Demilitarized Zone) with limited access to private network resources. The private, internal networks are separated from the DMZ with firewalls that apply rules that only allow connections to specific internal servers/services. VPN services to the internal network are protected with VPN management firewalls that apply network rules allowing only authorized and authenticated user traffic to traverse internally over an encrypted connection. Internal network subnets are separated based on access rules and allow traffic to egress to the Internet if it is authorized, as well as applying rules for network traversal between subnets. Finally, access to the innermost subnet is controlled by a firewall that allows only authorized traffic in or out, and only to specific hosts [18].

9.2.10.3 The Warmup

To allow our agents to train faster and in a more stable manner, we introduce a warmup phase prior to the training loop wherein our agents get to explore the network to record which nodes produce the largest positive reward upon discovery. This encourages our agents to access those nodes on its way to the target node in order to increase network discovery while maximizing positive reward. The steps of the warmup are as follows:

- We define a certain number of episodes for warmup phase in the training configuration.
- In the warmup episodes, the RL agent does not learn (no weight updates) but only monitors positive reward during an episode.
- If the RL agent receives a positive reward from a scanning action, then that node is added to the goal along with the target node and its reward for compromise is set to 100.
- The algorithm allows only one node (which gives out the maximum positive reward) from each subnet to be part of the goal as gaining control of one node in a subnet is enough to gain service information of the entire subnet.

Upon completion of the warmup phase, we now have a list of pathway nodes that the agent will attempt to discover on it's way to the target node in order to maximize network visibility. Note here than increasing the penalty factor will decrease the number of pathway nodes added to the target route. This lends itself to an intuitive explanation of risk. Much like taking a road trip, the higher the risk tolerance, the more you won't mind taking the scenic route on your way to the target. The lower your risk tolerance, the more you'll play it safe and get to where you need to go without a flat tire.

9.2.10.4 Results

Experimental results (Figure 9.1) show that training the dual agent model on our simulated network converges reliably for a variety of penalty factors. Furthermore,

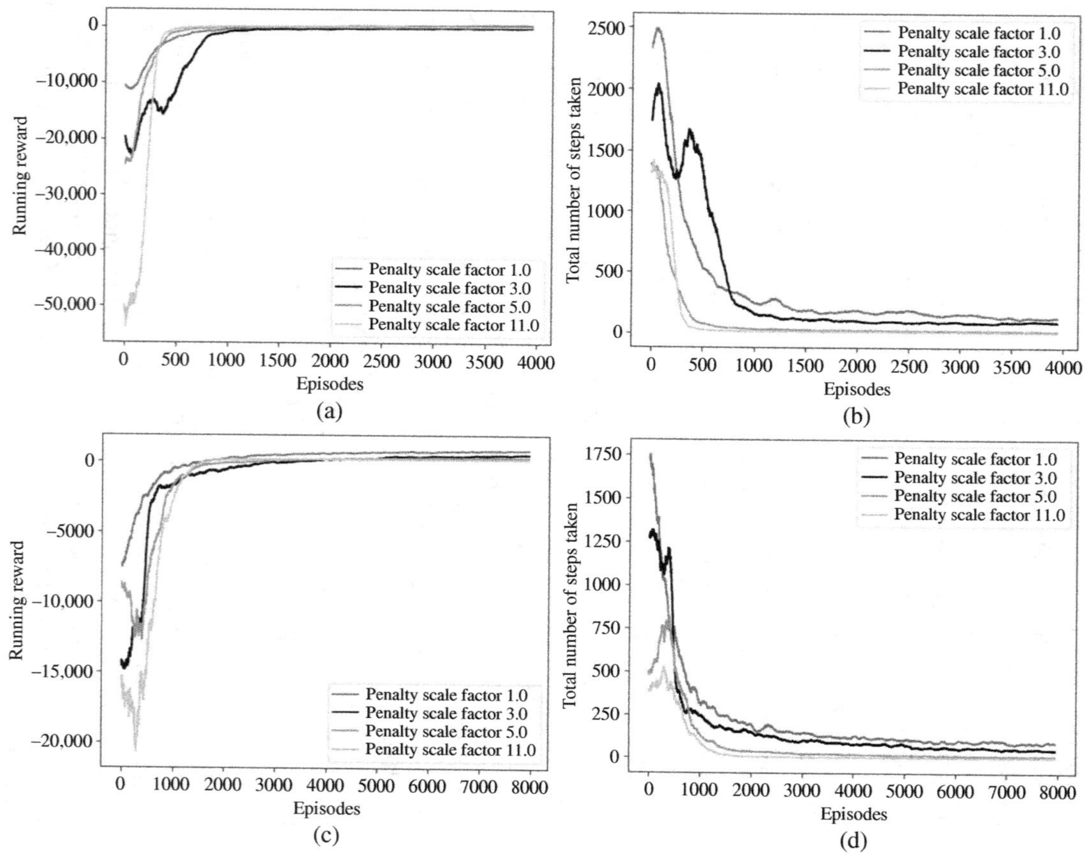

Figure 9.1 Training performance of DAA and A2C agents with different penalty scales.

a comparison of the PPO algorithm to traditional advantage actor critic (A2C) methods shows the PPO converges to an optimal solution faster than A2C for most penalty factors. More specifically, our dual agent PPO (DA-PPO) converges almost three times fast than DA-A2C methods. This implies that our dual agent PPO method learns and converges much faster than other methods and can more easily scale to enterprise networks without much performance degradation.

Furthermore, the number of hosts discovered by the agents on their path to the target node. Please note that the host identifiers in the above table are tuples representing the (*subnet*, *hostnumber*), respectively. We see here that for each target host, as the penalty factor increases, the number of hosts discovered by the agents consistently drops, representing decreased risk tolerance. Furthermore, we see that the agents learn to target a larger number of low penalty hosts than high penalty hosts without explicit knowledge of the services running on hosts. This shows that our simulations can realistically model cyber networks and behavior from an attacker perspective and produce intuitive results.

9.2.10.5 Creating Actionable Intelligence

We can use the output of the models and associated metadata to create actionable network topology diagrams that enable security operators to identify assets mostly like to be used during hostile surveillance.

Figure 9.2 shows a few such examples. Operators and enterprise security teams can use the output of these models for various proactive security measures such as enhanced monitoring of nodes frequently found on surveillance paths, patching, and reinforcement of particularly vulnerable hosts or subnets, or even dynamic network reconfiguration to make SDRs harder to discover for agents.

The possibilities with proactive cybersecurity are vast and hugely advantageous. Particularly, the ability to preemptively identify where an attacker would begin their campaign allows operators to mitigate risk before they become anything more. RL can enable this for security teams at scales never seen before while maintaining realistic decision making and risk vs. reward considerations.

9.2.10.6 Attack Surface Characterization (ASC)

Most attack surface characterization (ASC) descriptions focus on application vulnerabilities where tailored exploits can be weaponized to target these systems without considering the paths to traverse to take advantage of these weaknesses [22]. While acquiring a collection of discrete vulnerabilities and their relevant risk score is important to developing ASC, adversarial attack campaigns are in need of acquiring optimal avenues of approach. This optimization is based on weighing visibility considerations through maximizing the terrain visible while minimizing detection by deployed cybersecurity devices. In this book, we have

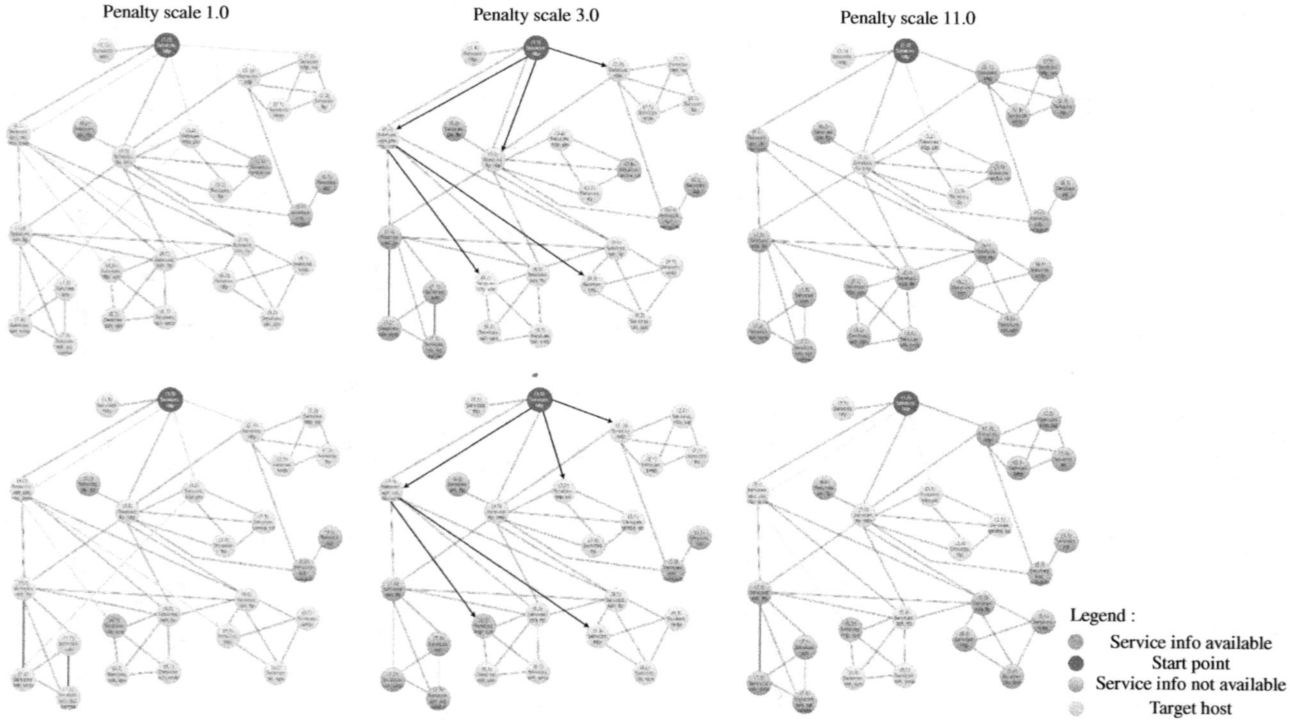

Figure 9.2 Network diagram showing the SDR for various penalty scale factors.

codified these visibility considerations into the rewards engineering of the RL for various models discussed in earlier chapters [6, 11, 18, 27].

This book widens the aperture by formulating ASC in adversarial terms by including the path analysis between applications deployed on systems on networks as recommendations from the RL trained on cyber terrain enriched attack graphs. Adversarial campaign planning leverages attack graphs as a critical ingredient to ASC and subsequent approaches for compromising enterprise infrastructures. Figure 9.3 shows how a network is translated to an attack graph where an RL model can predict the best paths for pursuing goals that align with attack campaigns or improving defensive posture.

ASC that combines application/system information with path traversal data is designed to support both red and blue workflows for mutual benefit. Such an ASC

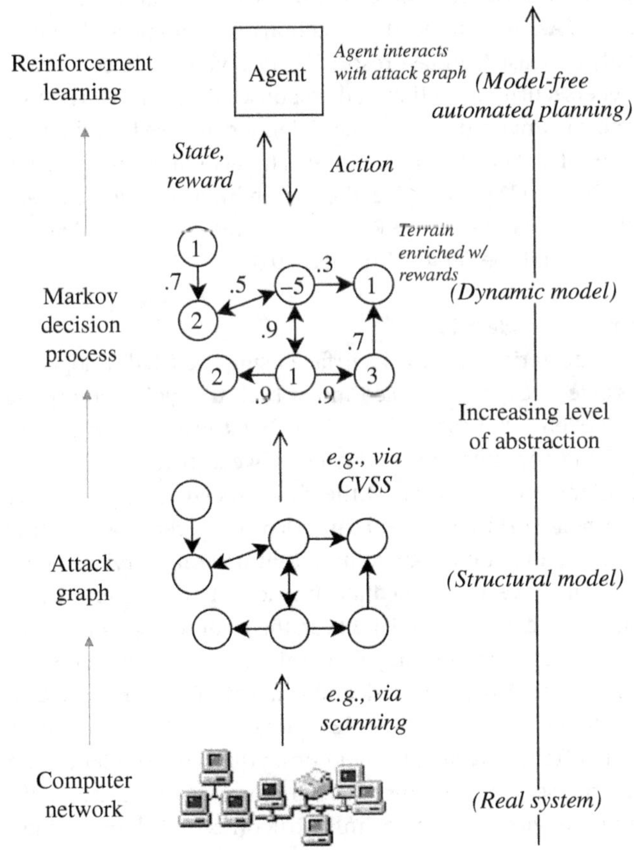

Figure 9.3 MDPS from attack graphs. Source: From Wang et al. [39], 2023, IEEE.

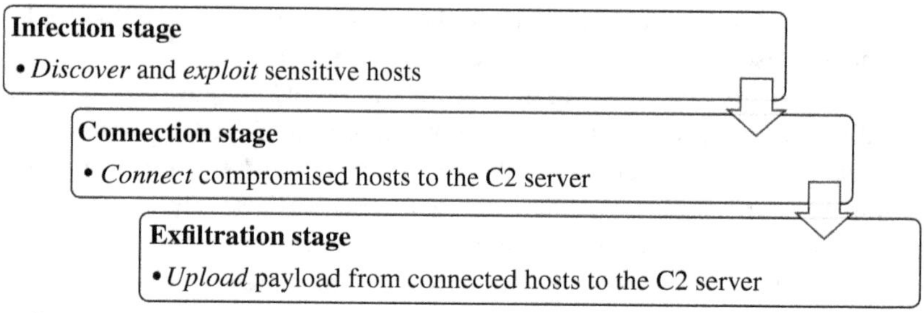

Figure 9.4 Overview of stages involved in the deployment of command and control (C2) infrastructure. Source: From Wang et al. [39], 2023, IEEE.

approach involves a nuanced understanding of optimizing visibility trade-offs by allowing the RL to model minimization of detection in the face of varying rewards engineering configurations. As an example, the detection of command and control (C2) channels is typically a measure of last resort for observing undetected malware that has evaded cybersecurity detection and response (D&R) infrastructure. Awareness of C2 channels is imperative for cyber defenders focused on the blue side to reduce the chances of dormant malware threats that could be awakened to execute nefarious tasks. From a C2 perspective, the overarching detection strategy is described in Figure 9.4. Upon applying the RL to a cyber terrain enriched attack graph, path analysis can be made as depicted in figure 9.5

9.2.10.7 Risk Management Considerations

With a view of how to characterize the attack surface using the RL developed in this book, and it is possible to express risk scoring at both the global enterprise and the local (sub) net/system/application level. This book provides around an InfoSec theme based on the common cause of gaps between true and assigned risks and values as being the misevaluation of the risk associated with visibility and network exposure. A real-world example of this is a print server. In evaluating a print server in CJA, an organization may assume that printing services being inoperable or degraded will have no immediate impact on the organization's business, nor impact other critical IT assets. However, the print server could serve as a pivot point[3] within the network as it may lie within a 1- or 2-hop graph of a target CJ. By inappropriately identifying the risk and value of a print server within close proximity to a CJ, the organization will underestimate its value in an adversarial attack campaign. This failure results from ignoring the need for identifying exploitable terrain (i.e., print servers and other infrastructure) within the breadth of the enterprise integrated services in complex infrastructures [8, 12]. In addition

3 Pivot points are footholds in networks for lateral movement.

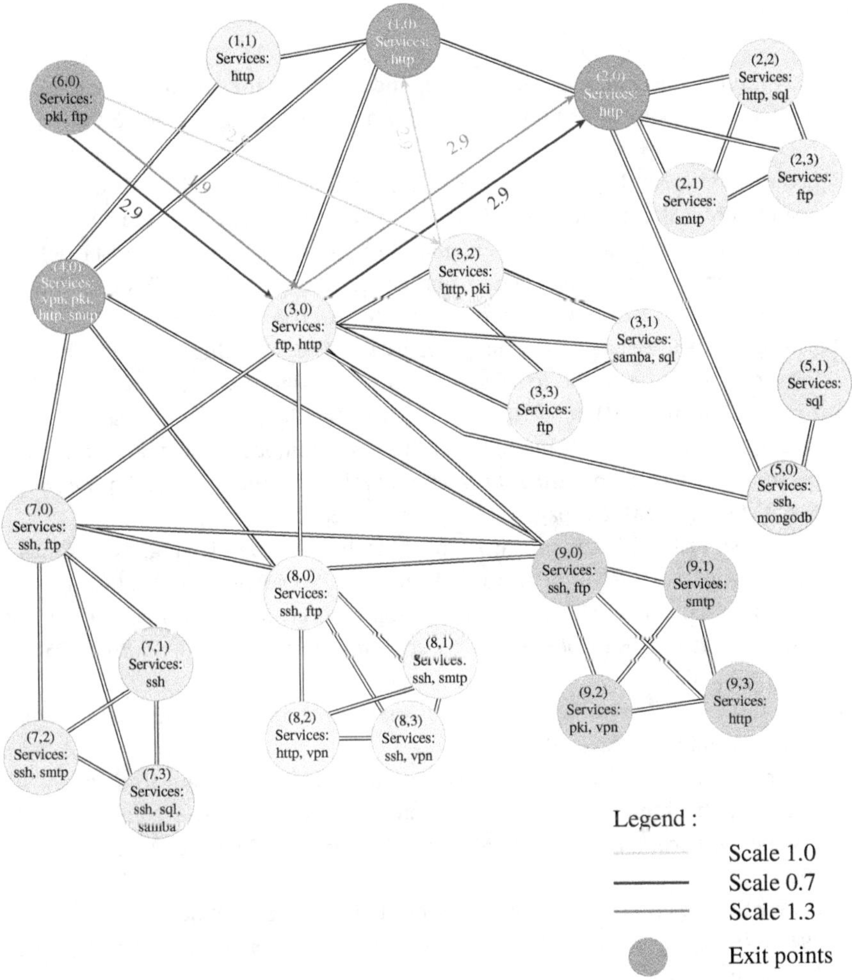

Figure 9.5 RL recommendation of best path to build for C2 channel based on visibility considerations. Source: From Wang et al. [39], 2023, IEEE.

to printing, users in the midst of their day-to-day tasks depend on services like authentication through identity and access management (IdAM), network access to key business shares, and connectivity to the internet from within the enterprise [7]. The integration of such services forms tertiary yet key aspects in the analysis of CJ that are critical to business function. The method in this paper, CJA-RL, offers a means of decreasing these tangential attack opportunities by identifying key terrain and avenues of approach and by identifying risks based on network structure and cyber terrain; contrary to most gap-prone threat-model-driven assessments.

References

1 Massimiliano Albanese, Sushil Jajodia, and Sridhar Venkatesan. Defending from stealthy botnets using moving target defenses. *IEEE Security & Privacy*, 16(1):92–97, 2018.

2 Scott Douglas Applegate, Christopher L Carpenter, and David C West. Searching for digital hilltops. *Joint Force Quarterly*, 84(1):18–23, 2017.

3 Sujita Chaudhary, Austin O'Brien, and Shengjie Xu. Automated post-breach penetration testing through reinforcement learning. In *2020 IEEE Conference on Communications and Network Security (CNS)*, pages 1–2. IEEE, 2020.

4 Chung-Kuan Chen, Zhi-Kai Zhang, Shan-Hsin Lee, and Shiuhpyng Shieh. Penetration testing in the IoT age. *Computer*, 51(4):82–85, 2018.

5 Ankur Chowdhary, Dijiang Huang, Jayasurya Sevalur Mahendran, Daniel Romo, Yuli Deng, and Abdulhakim Sabur. Autonomous security analysis and penetration testing. In *2020 16th International Conference on Mobility, Sensing and Networking (MSN)*, pages 508–515. IEEE, 2020.

6 Tyler Cody, Abdul Rahman, Christopher Redino, Lanxiao Huang, Ryan Clark, Akshay Kakkar, Deepak Kushwaha, Paul Park, Peter Beling, and Edward Bowen. Discovering exfiltration paths using reinforcement learning with attack graphs. In *2022 IEEE Conference on Dependable and Secure Computing (DSC)*, pages 1–8, 2022.

7 Greg Conti and David Raymond. *On cyber: towards an operational art for cyber conflict*. Kopidion Press, 2018.

8 Matthew Denis, Carlos Zena, and Thaier Hayajneh. Penetration testing: concepts, attack methods, and defense strategies. In *2016 IEEE Long Island Systems, Applications and Technology Conference (LISAT)*, pages 1–6. IEEE, 2016.

9 Laurent Gallon and Jean Jacques Bascou. Using CVSS in attack graphs. In *2011 6th International Conference on Availability, Reliability and Security*, pages 59–66. IEEE, 2011.

10 Rohit Gangupantulu, Tyler Cody, Abdul Rahman, Christopher Redino, Ryan Clark, and Paul Park. Crown jewels analysis using reinforcement learning with attack graphs. In *2021 IEEE Symposium Series on Computational Intelligence (SSCI)*, pages 1–6, 2021.

11 Rohit Gangupantulu, Tyler Cody, Paul Park, Abdul Rahman, Logan Eisenbeiser, Dan Radke, and Ryan Clark. Using cyber terrain in reinforcement learning for penetration testing. In *2022 IEEE International Conference on Omni-layer Intelligent Systems (COINS)*, pages 1–8, 2022.

12 Daniel Geer and John Harthorne. Penetration testing: a duet. In *18th Annual Computer Security Applications Conference, 2002. Proceedings*, pages 185–195. IEEE, 2002.

13 Mohamed C Ghanem and Thomas M Chen. Reinforcement learning for intelligent penetration testing. In *2018 2nd World Conference on Smart Trends in Systems, Security and Sustainability (WorldS4)*, pages 185–192. IEEE, 2018.

14 Mohamed C Ghanem and Thomas M Chen. Reinforcement learning for efficient network penetration testing. *Information*, 11(1):6, 2020.

15 Colleen Glenn, Dane Sterbentz, and Aaron Wright. Cyber threat and vulnerability analysis of the us electric sector. Technical report, Idaho National Lab. (INL), Idaho Falls, ID USA, 2016.

16 Jeffrey Guion and Mark Reith. Cyber terrain mission mapping: tools and methodologies. In *2017 International Conference on Cyber Conflict (CyCon U.S.)*, pages 105–111, 2017. doi: 10.1109/CYCONUS.2017.8167504.

17 Zhenguo Hu, Razvan Beuran, and Yasuo Tan. Automated penetration testing using deep reinforcement learning. In *2020 IEEE European Symposium on Security and Privacy Workshops (EuroS&PW)*, pages 2–10. IEEE, 2020.

18 Lanxiao Huang, Tyler Cody, Abdul Rahman, Christopher Redino, Ryan Clark, Akshay Kakkar, Deepak Kushwaha, Paul Park, Peter Beling, and Edward Bowen. Exposing surveillance detection routes via reinforcement learning, attack graphs, and cyber terrain. In *2022 21st IEEE International Conference on Machine Learning and Applications (ICMLA)*, pages 1–8, 2022.

19 Lanxiao Huang, Tyler Cody, Christopher Redino, Abdul Rahman, Cheng Wang, Ryan Clark, Edward Bowen, Peter Beling, and Ming Jin. Automated preference-based penetration testing with multi-objective reinforcement learning. *unpublished*, 2024.

20 Robert Larkin, Steven Jensen, Daniel Koranek, Barry Mullins, and Mark Reith. Towards dynamically shifting cyber terrain with software defined networking and moving target defense. In *International Conference on Cyber Warfare and Security*, pages 535–XIII. Academic Conferences International Limited, 2021.

21 Ryusei Maeda and Mamoru Mimura. Automating post-exploitation with deep reinforcement learning. *Computers & Security*, 100:102108, 2021.

22 Pratyusa K Manadhata and Jeannette M Wing. *A formal model for a system's attack surface*, pages 1–28. Springer New York, NY, 2011.

23 James P McDermott. Attack net penetration testing. In *Proceedings of the 2000 Workshop on New Security Paradigms*, pages 15–21, 2001.

24 Peter Mell, Karen Scarfone, and Sasha Romanosky. A complete guide to the common vulnerability scoring system version 2.0. In *Published by FIRST-forum of Incident Response and Security Teams*, volume 1, page 23, 2007.

25 MITRE. MITRE ATT&CK Framework, 2023. URL https://attack.mitre.org.

26 MITRE. MITRE Crown Jewels Analysis (CJA), 2024. URL https://www.mitre.org/publications/systems-engineering-guide/enterprise-engineering/systems-engineering-for-mission-assurance/crown-jewels-analysis.

27 Sathvik Murli, Dhruv Nandakumar, Prabhat Kushwaha, Cheng Wang, Christopher Redino, Abdul Rahman, Shalini Israni, Tarun Singh, and Edward Bowen. Cross-temporal detection of novel ransomware campaigns: a multi-modal alert approach. *arXiv:2309.00700 [cs.CR]*, 2023.

28 Scott Musman, Mike Tanner, Aaron Temin, Evan Elsaesser, and Lewis Loren. A systems engineering approach for crown jewels estimation and mission assurance decision making. In *2011 IEEE Symposium on Computational Intelligence in Cyber Security (CICS)*, pages 210–216, 2011. doi: 10.1109/CICYBS.2011.5949403.

29 National Institute of Standards and Technology. Security and privacy controls for federal information systems and organizations. Technical Report NIST Special Publication 800-53 Revision 5, U.S. Department of Commerce, Washington, DC, USA, 2020.

30 Hoang Viet Nguyen, Songpon Teerakanok, Atsuo Inomata, and Tetsutaro Uehara. The proposal of double agent architecture using actor-critic algorithm for penetration testing. In *ICISSP*, pages 440–449, 2021. doi: 10.5220/0010232504400449.

31 Xinming Ou, Sudhakar Govindavajhala, and Andrew W Appel. MulVAL: A logic-based network security analyzer. In *USENIX Security Symposium*, volume 8, pages 113–128. Baltimore, MD, 2005.

32 Joe R Reeder and Cadet Tommy Hall. Cybersecurity's pearl harbor moment: lessons learned from the colonial pipeline ransomware attack. *The Cyber Defense Review*, 6:15–40, 2021.

33 Robin Schoemaker, Rody Sandbrink, and Graeme van Voorthuijsen. Intelligent route surveillance. In Edward M. Carapezza, editor, *Unattended ground, sea, and air sensor technologies and applications XI*, volume 7333, pages 83–90. International Society for Optics and Photonics, SPIE, 2009.

34 John Schulman, Filip Wolski, Prafulla Dhariwal, Alec Radford, and Oleg Klimov. Proximal policy optimization algorithms. *arXiv preprint arXiv:1707.06347*, 2017.

35 Jonathon Schwartz and Hanna Kurniawati. Autonomous penetration testing using reinforcement learning. *arXiv preprint arXiv:1905.05965*, 2019.

36 Blake E Strom, Andy Applebaum, Doug P Miller, Kathryn C Nickels, Adam G Pennington, and Cody B Thomas. MITRE ATT&CK: design and philosophy. *Technical report*, 2018.

37 Stefan Varga, Joel Brynielsson, and Ulrik Franke. Cyber-threat perception and risk management in the Swedish financial sector. *Computers & Security*, 105:102239, 2021.

38 Sridhar Venkatesan, Massimiliano Albanese, Ankit Shah, Rajesh Ganesan, and Sushil Jajodia. Detecting stealthy botnets in a resource-constrained environment using reinforcement learning. In *Proceedings of the 2017 Workshop on Moving Target Defense*, pages 75–85, 2017.

39 Cheng Wang, Akshay Kakkar, Chris Redino, Abdul Rahman, S Ajinsyam, Ryan Clark, Daniel Radke, Tyler Cody, Lanxiao Huang, and Edward Bowen. Discovering command and control (C2) channels using reinforcement learning (RL). In *Submitted IEEE SouthEastCon 2023*, 2023.

40 Clark Weissman. Penetration testing. *Information Security: An Integrated Collection of Essays*, 6:269–296, 1995.

41 Mehdi Yousefi, Nhamo Mtetwa, Yan Zhang, and Huaglory Tianfield. A reinforcement learning approach for attack graph analysis. In *2018 17th IEEE International Conference on Trust, Security and Privacy in Computing and Communications/12th IEEE International Conference on Big Data Science and Engineering (TrustCom/BigDataSE)*, pages 212–217. IEEE, 2018.

A

Appendix

Common Top Internal Network Ports and Services
Port Number Service Description

- 22 SSH secure shell for secure remote access
- 80 HTTP hypertext transfer protocol (web browsing)
- 443 HTTPS HTTP over TLS/SSL (secure web browsing)
- 445 SMB server message block (Windows file sharing)
- 3389 RDP remote desktop protocol (Windows remote access)
- 53 DNS domain name system (name resolution)
- 139 NetBIOS network basic input/output system (legacy)
- 3306 MySQL MySQL database server
- 1433 MSSQL Microsoft SQL server
- 389 LDAP lightweight directory access protocol

Common Internet (Externally Observed) Ports and Services
Port Number Service Description

- 80 HTTP hypertext transfer protocol (web browsing)
- 443 HTTPS HTTP over TLS/SSL (secure web browsing)
- 53 DNS domain name system (name resolution)
- 21 FTP file transfer protocol (file transfer)
- 25 SMTP simple mail transfer protocol (email sending)
- 110 POP3 post office protocol version 3 (email retrieval)
- 143 IMAP internet message access protocol (email retrieval)
- 22 SSH secure shell for secure remote access
- 23 Telnet unencrypted remote access
- 3389 RDP remote desktop protocol (Windows remote access)

Reinforcement Learning for Cyber Operations: Applications of Artificial Intelligence for Penetration Testing, First Edition. Abdul Rahman, Christopher Redino, Dhruv Nandakumar, Tyler Cody, Sachin Shetty, and Dan Radke.

Index

Reinforcement Learning for Cyber Operations: Applications of Artificial Intelligence for Penetration Testing, First Edition. Abdul Rahman, Christopher Redino, Dhruv Nandakumar, Tyler Cody, Sachin Shetty, and Dan Radke.

www.ingramcontent.com/pod-product-compliance
Lightning Source LLC
LaVergne TN
LVHW021127130325
805789LV00001B/19